AF887376

Mysteries *of the* Universe

(The Unseen)

Mysteries *of the* Universe

(The Unseen)

Deepankar Singh

Mysteries of The Universe – The Unseen
Deepankar Singh

Absolute Author
Publishing House

All rights reserved
First Edition, 2020
© Deepankar Singh, 2020

This book contains material protected under International and Federal Copyright Laws and Treaties. Any unauthorized reprint or use of this material is prohibited. No part of this book may be reproduced or transmitted in any form or by any means, electronic or mechanical, including photocopying, recording, or by any information storage and retrieval system without express written permission from the author/publisher.

Contact the author at astro.physics0110@gmail.com

Paperback ISBN: 978-1-64953-129-2
eBook ISBN: 978-1-64953-130-8

"A good philosopher should remind himself, now and then, that he is a particle pontificating on infinity."

TABLE OF CONTENTS

DEDICATION

To Lovers of Physics

I hope this book gives you a great understanding of tiny particles around us. I intend my research helps you - the reader – approach the field of particle physics from a new angle.

To My Parents

Thank you for always being unbiased support and raising me to be the person I am today.

To My Brother 'Himanshu'

Thank you for being a person who always says, 'let's see how we can fix this' instead of, 'I told you so.'

To My Loving Son 'Rudra'

This book was written in the first six months of your life. So, it will take a long time for you to make sense of all things written in it.

To My Precious Wife 'Ektaa'

Without your love and encouragement, this book might never have come to fruition. A special thanks to you for all your support and for always motivating me on my thoughts. Your commitments made it possible.

In memory of my uncle
Sunil Kumar.
You left fingerprints of grace on our lives.
You will always be in our hearts,
and our minds.
You shan't be forgotten

"What we know is a drop; what we don't know is an ocean."

Isaac Newton

PREFACE

What is the Universe?

Whenever we try to find the answer to this question, the first thing which comes to our mind is, "What is the building block of this universe? "in simple terms, "What is the universe made up of?"

If a school student answers this question, he will say that universe consists of **ATOMs**, which is the smallest thing in a universe with a nucleus in its center containing neutrons and protons inside it. Apart from these, there are Electrons present in an atom which revolves around the nucleus.

Suppose you ask the same question from a university student. In this case, he will go a bit further and say that Neutrons and Protons presented in the nucleus are not the elementary particles of an atom because they are made up of **QUARKS**.

By this conclusion, are we convinced that the **UNIVERSE** is an amalgamation of Electrons and Quarks and nothing else? Is the knowledge gained from the university enough to understand the **UNIVERSE**? The answer is **NO**.

The University model didn't explain to us how different forces and particles work.

This book will help you to uncover the astonishing and mind-blowing facts about the universe. And will give you a better understanding of particles and the theories behind them.

Let's start our journey with The Standard Model of Elementary Particles.

***

"Behind every atom of this world, hides an infinite universe."

Rumi

1

STANDARD MODEL OF ELEMENTARY PARTICLES

There was a time when we thought that an ATOM is the only tiniest particle of the matter in the universe, and we cannot break it more than a limit.

Gradually, we grasp that the atom is not the tiniest particle because it comprises Protons and Neutrons. Later, new and anonymous subatomic particles were discovered with the help of Quantum Theory.

Then, scientists needed a new theory to understand these new subatomic particles, and when they successfully made it, they predict the presence of several more subatomic particles.

A team of scientists then decided to put the charged particles in a particle accelerator and propel them on high energy.

Particle Accelerators produce beams of charged particles that can be used for a variety of research purposes. As a result, many particles colliding gave birth to many new, more particles, and the process goes on and on.

Due to many new particles, they did the subdivision of particles in different categories according to their properties. These properties are shown in Box 1.1.

These scientists designed a model that allows for adjusting these particles and improved understanding of how different forces work.

This model is called the **"Standard Model of Elementary Particles."** *(Image shown on page 4)*

Box 1.1: Properties of Particles

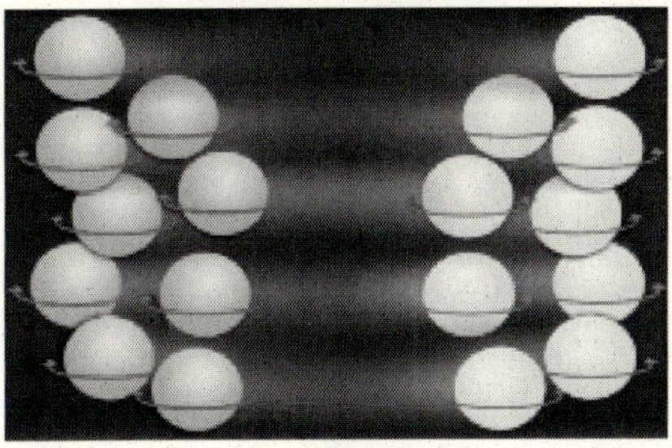

A. SPIN

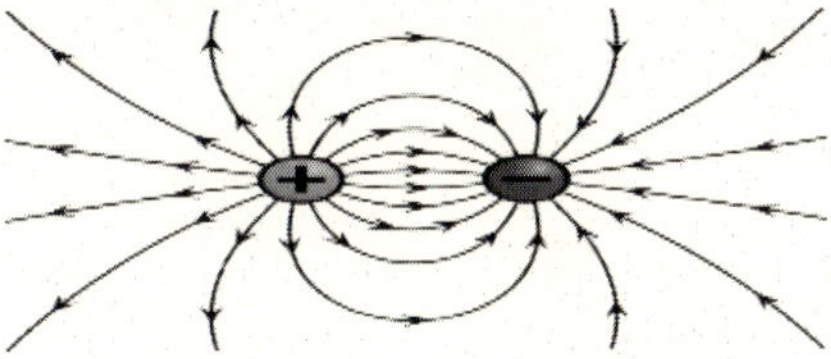

B. ELECTRICAL CHARGES

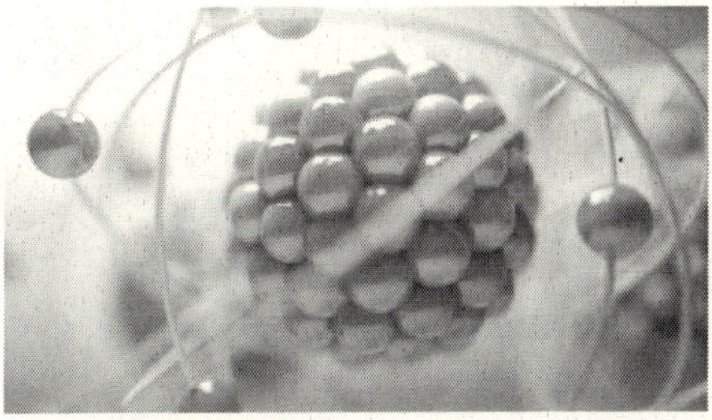

C. MASS

D. LIFETIME

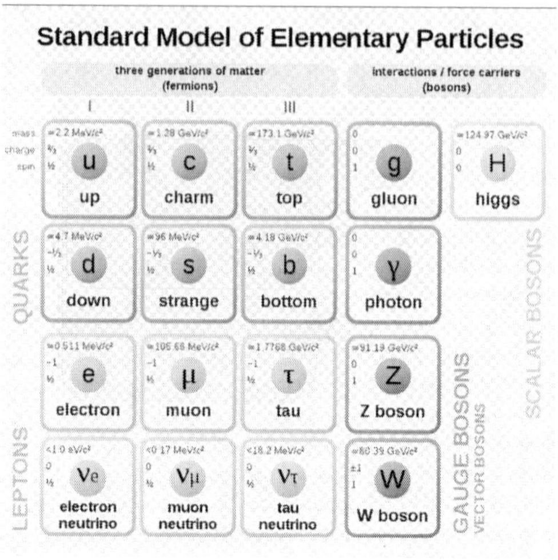

Let's take a deep dive into this model and get to know the roles of different particles.

All the particles present in the universe are divided into two categories, namely:

1. FERMIONS and
2. BOSONS

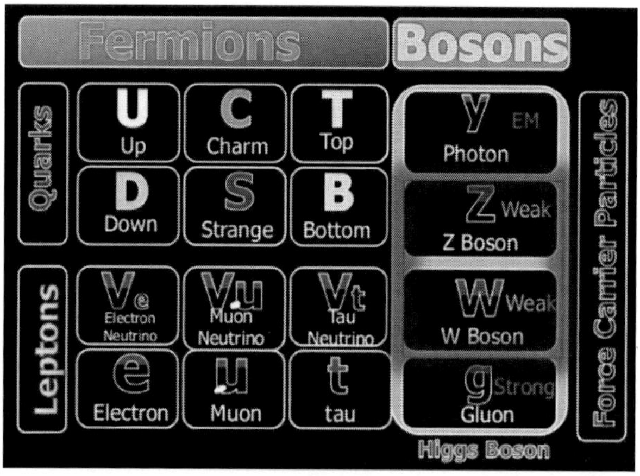

1.1 - The 'Fermion'ization.

Fermions are responsible for customizing every form of "matter" in the universe, whether it's a tree, a mountain, a planet, or anything else. This family contains two more kinds of particles named Quarks and Leptons.

These are the ones responsible for the creation of Proton and Neutron inside the nucleus. Quarks are never stable; that is why they combine and construct the new particles called **HADRONS**.

Hadrons also divided into two particles

BARYONS = 3 Quarks (protons and neutron)
MESONS = 1 Quark + 1 Anti Quark (pion, kaon)

FLAVORS OF QUARKS: Up(U), Down(D), Charm(C), Strange(S), Top(T), Bottom(B).

Protons and Neutrons are also a part of Baryons as they consist of a combination of Quarks.

The remaining 4 Quarks (Charm (c), Straight (S), Top (T), Bottom (B)) combines and creates different types of combinations, which results in the discovery of new kind of particles.

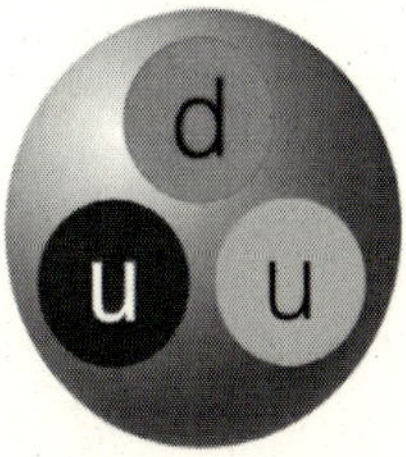

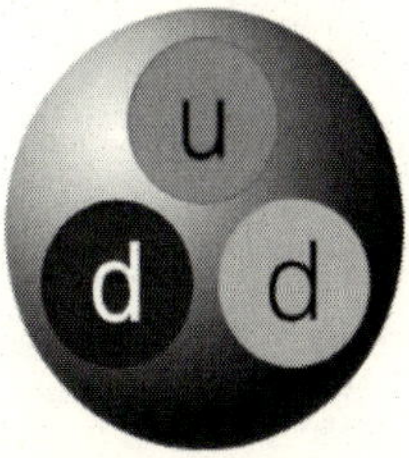

1.2 - Electrons with 'colleagues.'

1.2.1 Leptons

Back to our discussion on Leptons. These are the particles that are NOT made up of Quarks.

PROTONS = 2 UP quarks and 1 DOWN quark (UUD)

NEUTRONS = 2 DOWN quarks and 1 UP quark (UDD)

This family consists of *Electrons, Muon, Tau, and Neutrino*.

These are stable particles because they do not combine within themselves.

"The Heisenberg Uncertainty Principle" predicts those particles which exists in the universe without any real cause because of the probability of their existence. Hence, they are known as ***Virtual Particles***. We will cover in-depth The Heisenberg Uncertainty Principle in another part.

$$\Delta x \Delta p \geq \frac{\hbar}{2}$$

Mathematical representation of Heisenberg's Uncertainty Principle

Stay focused! We are still discussing Leptons.

Leptons are *Virtual Particles* because of the probability of the existence of Leptons by most scientists in the universe. Irrespective of their quaint behavior, they follow a specific pattern.

Surprisingly, they appear in a pair of one Regular Matter and one Anti Matter. This appearance is the reason that an Electron always constructs with its Anti Matter partner known as Positron.

The mass of both Electron and Positron is the same; however, the charge is opposite. Because it built up energy itself when they collide, they emit a high amount of pure energy as an output.

Like Electrons, the other Leptons also have their Anti-Matter. Now it is clear that only Fermions are responsible for every kind of matter created in the universe. But they have no control over how forces works.

BOSONS do this work.

1.3 - The Transporter

The name, Boson, was coined by Mr. Paul Dirac to commemorate the contribution of Mr. Satyendra Nath Bose, who was an Indian Physicist and developed Bose-Einstein Statistics, which theorizes the characteristics of elementary particles.

There are four types of BOSONS, namely:

1. PHOTONS
2. W/Z Boson
3. Gluons
4. Gravitons

In older times, the Electrical Force and Magnetic Forces were the two different forces available.

But now we consider them as the one force called **"ELECTROMAGNETIC FORCE"** or **Lorentz Force**, which combines both electricity and magnetism manifestations of the same phenomenon.

This force works because of **PHOTONS**. Without photons in Electromagnetic force, a magnet cannot attract the Iron, and neither the electricity will work.

We will not stretch this topic because you are already used to and experience 'Electro-Magnetic Forces' in your daily lives.

From all four fundamental forces, 'Weak Nuclear Force' is considered as the second weakest force. This weakness is the reason Beta Decay takes place, and it will create Anti Neutrino.

W/Z BOSONS are the carrier of this force.

'Strong Nuclear Force' is the strongest of all the four forces. This great strength is the reason why Protons and Neutrons stick together in the nucleus.

GLUONS work as a transporter of this force.

The weakest force is the 'Gravitational Force.' Unfortunately, we didn't find any particle responsible for Gravitational Force on the Quantum level until now. But there are anticipations that there will be an existence of this kind of particle.

For now, this hypothetical particle is called **GRAVITONS**.

The only thing we didn't go through on this topic is **HIGGS BOSON**. We will try to understand this particle in another chapter. For now, just keep in mind that Higgs Boson is the particle that gives mass to all other particles.

You already know about particles that make the matter and those particles that work as the forces' carriers.

However, according to the Quantum Field Theory, this universe's building blocks are not Quarks and Electrons.

You are surprised, right?

According to the Quantum Field Theory, if we provide the energy to the fields, it will originate the particles like quarks and electrons. We will go in-depth with 'Quantum Field Theory' in another part of this book.

For now, we will try to understand about **HIGGS BOSON** also known as **"THE GOD PARTICLE."**

Things You Need To Know

- Quarks and Leptons are the particles from which matter is created.

- Bosons are particles that work as a transporter of all fundamental forces.

- Going deeper in Standards Model, the result for the numbers gotten are: 48 Matter particles containing Leptons and Quarks, and 12 Forces contain 8 Gluons, W+, W-, and Z (Carriers of the weak force) and Photons.

Interesting Fact

The word "ATOM" comes from the Ancient Greek word meaning "UNDIVIDED". They thought that an atom could not be cut into anything smaller, which we know is not exactly right.

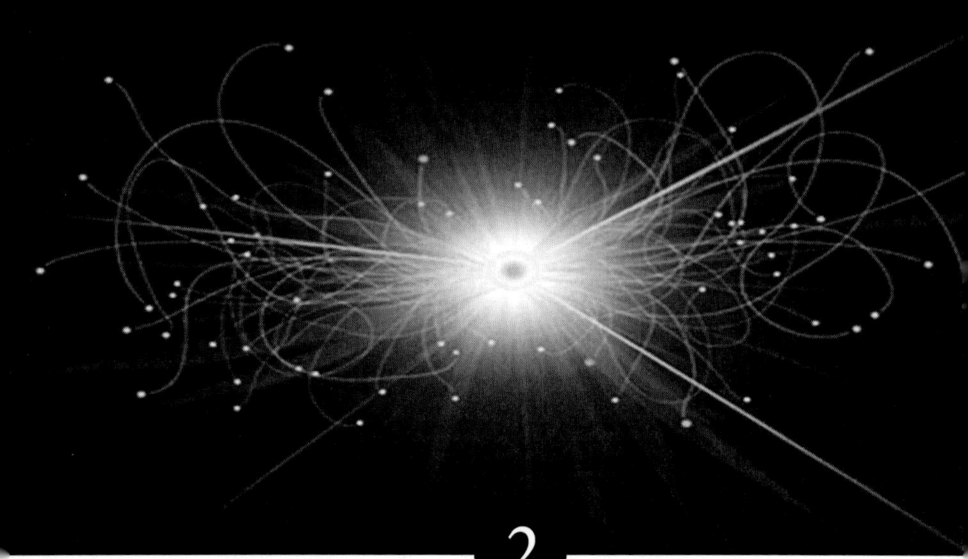

2

THE GOD PARTICLE

"Higgs Boson," is also known as The God Particle.

You might have a question in your mind that why it's called a God Particle.

Right?

It's correct to say that the whole anecdote of the Higgs Boson started due to the mass.

So, to understand Higgs Boson, we first need to understand that what is MASS?

In simple terms, mass is the measurement of the inertia of a body.

When trying to change the state of any physical body, either at rest or in uniform motion, you will feel a resistance from that physical body. That resistance is called Mass.

For example, when you are trying to push a stopped car to move it in a forward direction, you are experiencing its mass.

If you still have confusion with the understanding of mass, we will try to make it simpler because we must grasp mass to understand Higgs Boson.

2.1 - Mass or Weight?

We know that anything that does not have its mass can travel at the speed of light in a vacuum.

PHOTONS are an excellent example of this phenomenon.

This means that anything that is not traveling at the speed of light must have some mass.

Before we move further, I will make sure that you will not confuse Mass with Weight.

Yes, Mass and Weight are related to each other but are entirely different.

Mass vs. Weight

Mass and Weight are *RELATED*
but are *NOT* the *SAME* properties.

Weight contains the contribution of gravity in it. However, the mass has nothing to do with gravity.

That is why the mass of anybody will be the same in any part of the universe.

However, its weight will fluctuate on different planets. It will increase or decrease according to the gravity of that planet. (See picture illustration below)

We know that everything in this universe has its mass. But what are their sources?

You will probably say that it's made up of its constituent particles OR the particles from which the thing is made up.

For example, A mass of a car is a combined mass of its different parts.

But if we analyze at Quantum Level, we see that Leptons and Quarks are the ONLY building blocks of the matter, which means that they are the smallest particle and can't be broken down any further.

These are fundamental particles and not made up of any other thing, and these should be considered a completely massless particle.

Right? But this is far from the ground reality!

As we all know, Electrons have mass; likewise, Quarks also have mass in a particular manner.

So, the question is:

What are the sources of the masses??

What is that unusual thing in the universe that is providing mass to these particles?

If Quarks and Electrons contain the mass in them, then why don't Photons have one?

2.2 - The Scepticism

In the 1960s, a team of physicists composed few mathematical equations based on elementary particles' behavior.

They assumed that these equations would help them to get an in-depth understanding of the elementary particles.

But those equations pestered them even more.

Because whenever they solve the equations assuming that all the elementary particles like Electrons and Quarks are massless, the equations are solving correctly, and the predictions made based on those equations were also getting 100% accuracy, which was experimentally tested and verified too.

Physicists were already aware that Electrons and Quarks are not massless particles, and they carry some mass.

As per this scenario, whenever they try to solve the equation by putting an assumed mass to the particles, the same equation gets more complex and inconsistent.

Physicists failed to understand the reason behind it and what might be the solution for this.

It is time Mr. Peter Higgs stepped in and proposed a new theory in which he predicts the existence of Higgs Field and Higgs Boson.

Photo of Mr. Peter Higgs

According to The Higgs Theory:
Everywhere, and every period of this universe, consists of different Fields and not the particles. Fields like Electron Field, Quark Field, Electric Field, Gravitational Field, etc. which are impossible to visualize.

The creation of particles takes place when we impose the energy into these fields. And the extraction of energy from these fields cause the annihilation of the particles.

For example: If we impose energy in Electron Field, then it will create the Electrons, and Quarks will be produced if we impose energy into Quark Field.

Similarly, every other field creates more other particles and interacts with each other as well.

According to Mr. Peter Higgs, there is one more field in the whole universe called "HIGGS FIELD," which is identical to all other fields.

When we impose energy to Higgs Field, it will give birth to a new particle HIGGS BOSON. (See picture representation below)

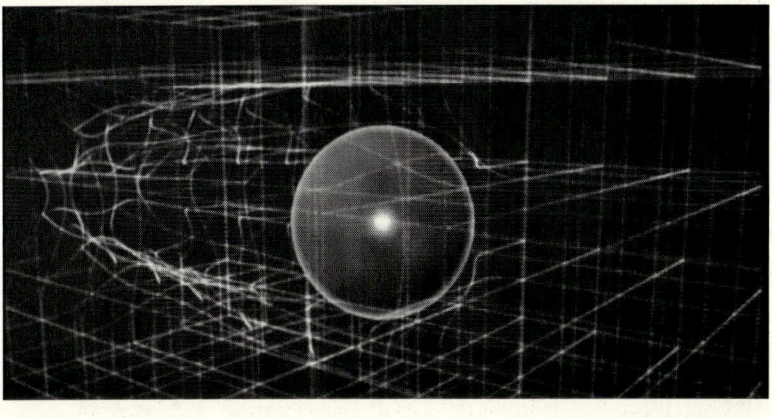

Mr. Peter Higgs follows a theory that all the particles are massless like Photons and it is the Higgs Field that provides them with mass.

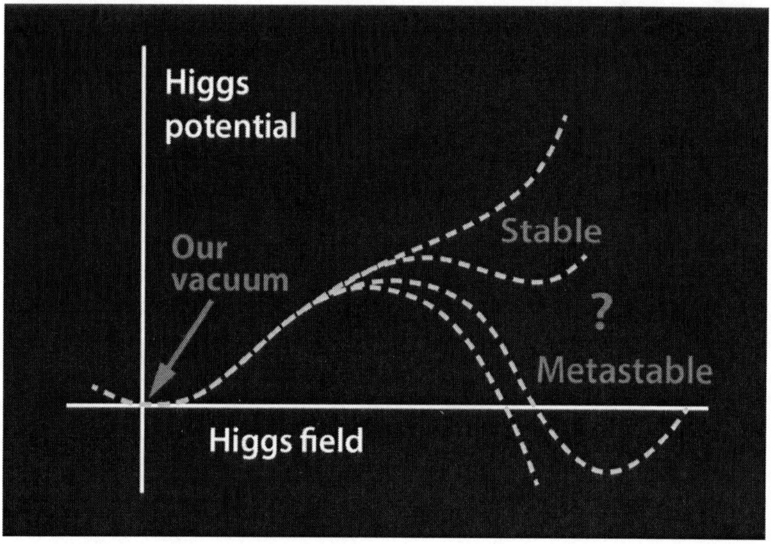

Now, you might be thinking about how this whole process works.

2.3 - The Higgs Mechanism

Let's go through the whole mechanism of Higgs Field.

According to Mr. Higgs Theory, all the elementary particles like Electrons, Quarks, and Photons are absolutely massless before entering into the Higgs Field.

However, after entering Higgs Field, some particles interact more with the field, and some interact significantly less.

The particles which interact more with the field get more mass. And, the particles which interact less get less mass.

For example, photons hardly interact with the field; that is why they remain massless.

Electrons interact less with the Higgs Field, so they have smaller mass.

Quarks interact more; that is why they have a larger mass.

THE HIGGS MECHANISM

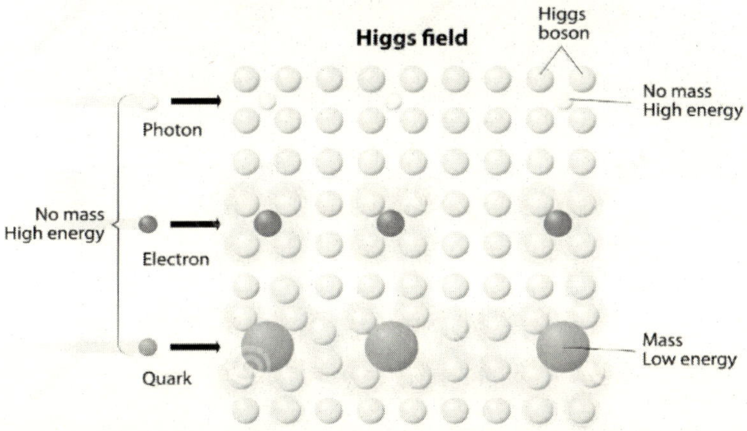

If you still have some problems understanding the Higgs Mechanism, let's discuss it with another straightforward example using the Bar Counter approach.

2.4 - The Bar Counter

Let's assume that there is a big party going on in a big hall. All invited people are physicist by profession, and everyone is discussing something interesting about physics with each other.

If an accountant enters a party, then no one would be interested in interacting with him. And, he will easily be able to reach the bar counter without any restrictions.

However, if Mr. Albert Einstein would enter the same party hall, every physicist presented there would want to speak or meet with him.

As a result, he will be surrounded by many people at the party in few minutes, and it will take a lot more time for reach him to the bar counter.

In the above example, let's assume all the physicists presented there as Higgs Field.

Assume the accountant as PHOTONS and Mr. Albert Einstein as QUARKS.

That means that the particles which have more interactions with Higgs Field will have more mass.

2.5 - Atom Smasher

In 1964, Mr. Peter Higgs published this theory through a famous mathematical journal, which was denied by everyone at that time.

Not because of some miscalculations in his mathematical equations, but because no one wanted to accept that the invisible fields can also give the mass to the particles.

But Mr. Higgs didn't give up on his theory and published the same in another physics journal.

The physicists that studied this theory, in depth, were very much convinced about this theory and were expecting the existence of Higgs Field based on the mathematical equations.

Even with the mathematical equations, it was difficult to reach any conclusions, and this field can only be detected experimentally.

From here, the world's most massive machine was invented and named the *Large Hadron Collider*.

The problem in identifying the existence of these fields is their invisibility. That means its impossible to detect them, as they are entirely invisible.

Mathematics taught us that, if our calculations are correct and Higgs Field exist somewhere in the universe, also, the collision of two charged particles at very high speed in a compact space can create a fuss in the Higgs Field for some time. This is analogous to a situation where two boats plow into each other on the water, the water creates a fuss in it.

The fuss created in a Higgs Field gives birth to the Higgs Boson particle, which can be seen and detected as well. This means that if we can detect Higgs Boson Particle, then we will discover the existence of Higgs Field in the universe.

2.6 - The Discovery

Following this theory, a team of physicists and scientists decided to collide the charged particles in the Large Hadron Collider by 99.99% speed of light.

They tried to find a decay pattern based on a mathematical equation associated with Higgs Boson.

However, they were unable to find anything from various experiments conducted till the 4th of July 2012. Then, CERN announced in the news that Higgs Boson Particle has been successfully detected.

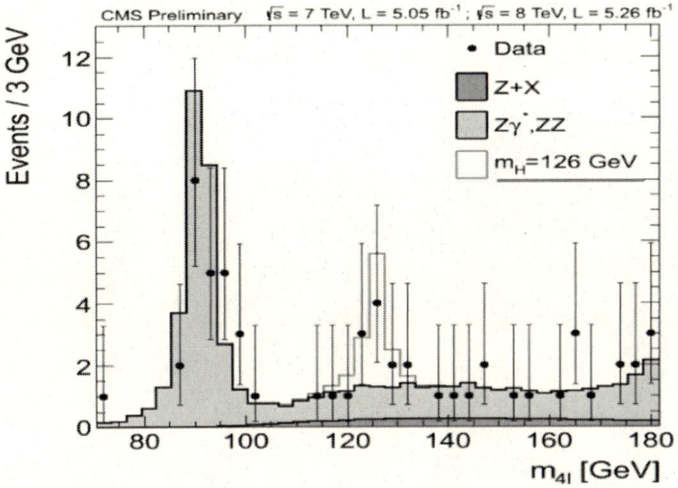

This discovery of a Higgs Boson Particle appears to manifest the theory of Mr. Peter Higgs.

Moreover, proof that his theory was 100% correct for which he was dignified by a Nobel prize in the year 2013.

Hopefully, now you have a clearer picture of HIGGS BOSON.

Some people have a misbelief that Higgs Boson provides mass to all elementary particles.

But let me clarify that Higgs Field is responsible for the mass of elementary particles and not Higgs Boson.

Higgs Bosons are just the particles constructed from the Higgs Field.

Now, you might have a few questions in your mind like:

Why was Higgs Boson named as a GOD PARTICLE?

Why are these Higgs Field vital for us?

Most people think that Higgs Boson was called a God particle because somehow it is related to a god.

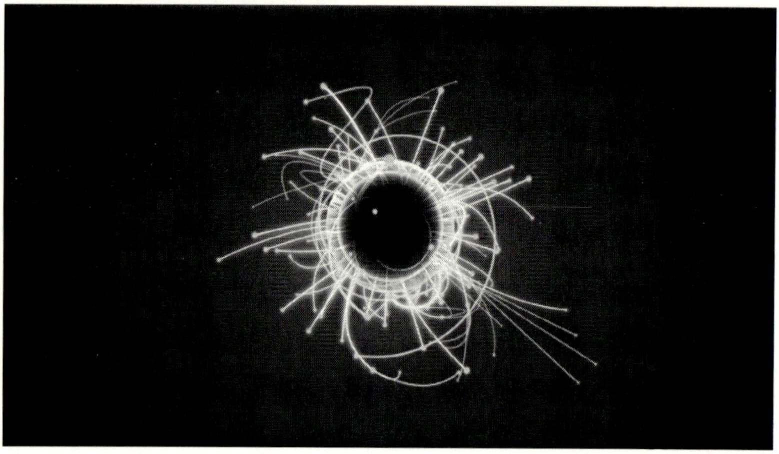

But this is completely erroneous information. As previously mentioned, it's Higgs Field and not the Higgs Boson that provides mass to the elementary particles.

If there would be no Higgs Field existed in the universe, then all the elementary particles would remain massless and travel at the speed of light. Due to this speed, neither electrons will revolve around the nucleus nor neutrons, and protons will always exist in the nucleus.

In other words, there will be no atoms, no molecules, no matter, or anything else existing in the universe.

In other words, there would be **NO UNIVERSE**.

There is an interesting story behind the name God Particle.

In the early '90s, a famous physicist, Mr. Leon Lederman, wrote a book on the same subject.

At first, he wanted to name that book "THE GODDAMN PARTICLE." But the publisher was getting skeptical about this name and decided to change it to "THE GOD PARTICLE."

The publisher had no logic behind the new name, and for them, it was just a name which could be more presentable.

However, the media had created so much hype of this name that it became very famous. Everyone forgot about the name Higgs Boson and started calling it the God Particle.

There are many reasons that the invention of Higgs Field considered a breakthrough in physics.

First Reason:

It is very non-identical with the other fields which are presented in the universe.

We know that there are many invisible fields like the magnetic field, gravitational field, electromagnetic field, etc. which existed in the space.

However, the effect of these forces is heterogeneous. That means that their nature varies in different places.

But Higgs Field is the only one whose nature is homogenous.

That means that it behaves in the same way in the universe. Whether it's your room or any other galaxy of the universe, the effect will be the same.

Scientists are trying to deeply understand the Higgs Field because it can help us solve many mysteries, including Dark Matter and Dark Energy.

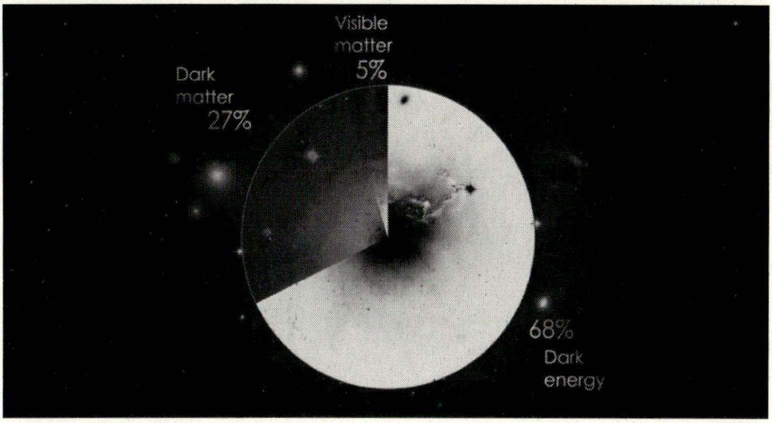

Second Reason:

We discovered a new conformation of the matter in the form of Higgs Bosons. Previously, these structures were only limited to a mathematical equation.

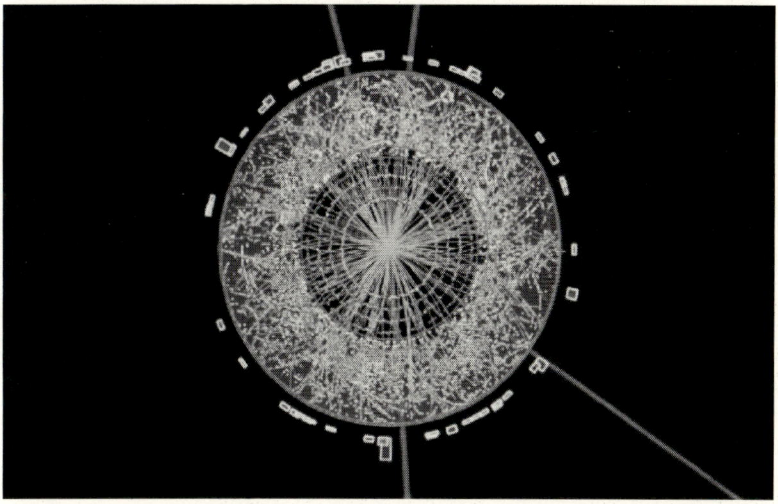

In a particle, except for its mass and charge, spin is the only thing that never changes. Its speed of a spin never increases nor decreases with the time.

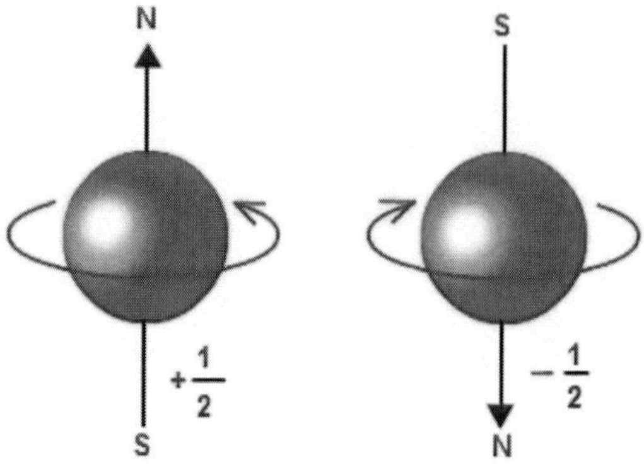

For Example, Electrons and Quarks have the same spin values. Photons have twice the spin value of Electrons and Quarks.

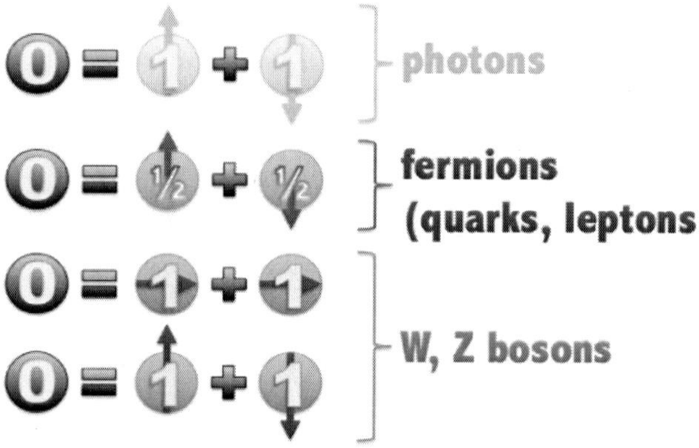

However, Higgs Boson is the only particle that has no spin at all.

Third Reason:

Higgs Field tells us, with the help of a mathematical equation, we can understand the nature of the universe and make accurate predictions of unknown things that are never experimentally detected.

Higgs Field has been experimentally detected in the year 2012 when we found Higgs Boson. However, it was predicted many years ago by a mathematical equation.

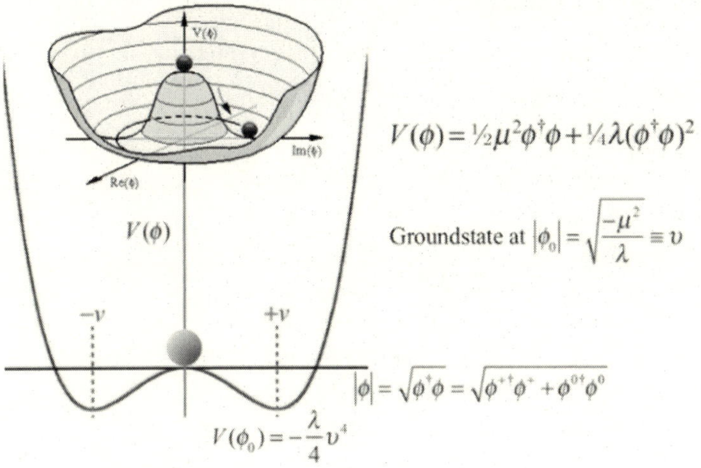

Moreover, some scientists disagree that the mathematical equation also predicted Higgs Boson particle. This was proved later by the Large Hardon Collider experiment.

Some people have a misconception that Higgs Boson is the smallest particle of all.

But it is very important to know that, according to the standard model, Higgs Boson particle decays very quickly into other particles and particle detectors are unable to detect it because of its size This means that Higgs Boson is still unknown.

This is the reason we cannot even call them true particle-like electrons. However, the mass of Higgs Boson is mostly higher than the Bosons.

Mass and Compton wavelength of Higgs Boson would be as below:

Mass = 125 Mev
Compton Wavelength = 10 ^-15cm
Comparison to which the Compton wavelength of Proton is:
Compton Wavelength = 10 ^-13cm

Now you might have a question that, who gives mass to Higgs Boson?

The answer is **HIGGS FIELD**.

The reason is the un-identical behavior of Higgs Field in comparison to other fields.

Only a few excitation states in all other fields are identified as particles, and the rest are considered zero value.

However, in Higgs Field, there is high value everywhere, but this high value is not the Higgs Boson.

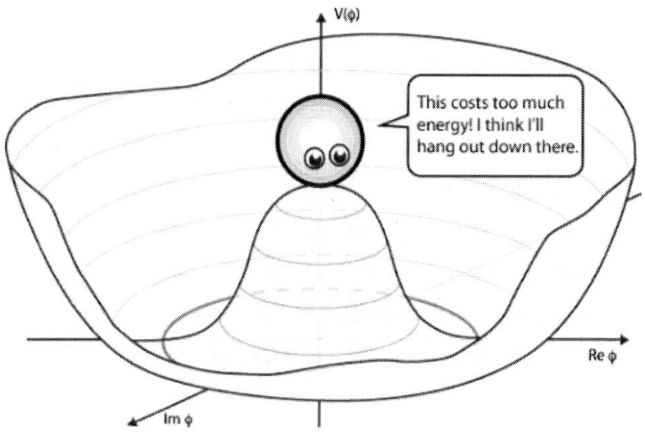

A field needs an extra excitation to produce a Higgs Boson Particle. And because there is a non-zero value everywhere in a Higgs Field, Higgs Boson starts interacting with its own field after producing and gaining its own mass.

That is how the Higgs Mechanism works and gives birth to Higgs Boson, aka *THE GOD PARTICLE.*

If GOD exists, then why not GHOSTS?

Yes, if there is a GOD PARTICLE present in this universe, then there is an existence of The GHOST PARTICLE as well.

So, let's continue the journey towards NEUTRINO - "The Ghost Particle."

The Least You Need To Know

- Weight and mass are related but are not the same.
- The God Particle has nothing to do with 'GOD.'
- Higgs Bosons are just the particles constructed from the Higgs Field.
- Higgs Field is responsible for the mass of elementary particle.

Interesting Fact

In Particle Physics, the term Quarkonium represents a flavorless meson whose constituent particles are a heavy quark and its own antiquark. This composition makes Quarkonium a neutral particle and the anti-particle of itself.

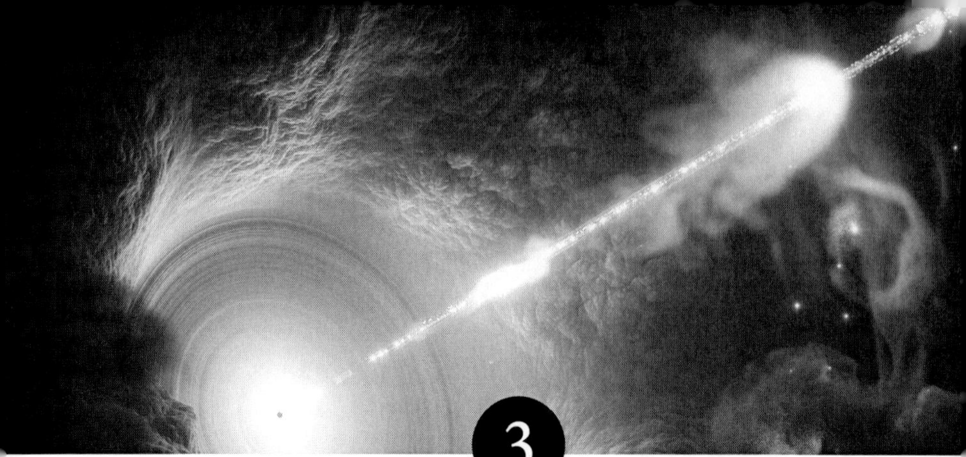

3

THE GHOST PARTICLE

As we all know that our universe is hiding many magical things and mysteries inside it, and moving along with time, we will discover new and unique things about it.

However, after trying our best to understand the universe, we still lack behind on a lot of facts.

Like the famous quote that says: "*What we know is just a drop; what we don't know is an ocean.*"

Image of a man thinking about the universe

Whenever we think that we already discovered a lot and start to believe that we know everything about the universe, it always comes up with more shocking and mind-blowing interesting things.

NEUTRINO, one of the mysterious fundamental particles, also known as The Ghost Particle, is just a tiny joint of a long chain.

Let's start with the question: What is Neutrino?

In a simple language, Neutrino is a fundamental particle about which we know the least.

Now, some people might want to ask that, what are these fundamental particles?

I would like to describe Fundamental Particles here so that it will be easier for you to understand things related to it.

Fundamental particles are those particles which are the tiniest in the universe. This means from an atom to a galaxy, every structure and every object existing in this universe is made up of fundamental particles.

3.1 - Quantum Realm

All the forces and phenomena presented in the universe are also connected to the fundamental particles.

That means fundamental particles are the basic building blocks of this universe.

If you are too curious to know about their size, then I would like to mention here about the movie "Ant-Man."

The main character (Scott Lang) in this movie can decrease his size to the size of an ant with his suit's help.

Due to some problems with his suit, he cannot control his size and get smaller and smaller infinitely at the end of the movie.

In this process, his size gets smaller than fundamental particles. The things which are smaller than the fundamental particles are named Quantum Realm in this movie.

Of course, it's a sci-fi movie and not the reality, but this gives an idea about the size of the fundamental particles compared to the ordinary objects.

When we talk about the fundamental particles, we only think about electrons, protons, and neutrons because we have studied mostly about these particles in school and university.

But the truth is, there are many more fundamental particles, and their combinations existed.

"Standard Model of Elementary Particles" gives all the information about all the particles and their combinations.

We have already covered it thoroughly in our first chapter.

However, let's go through it briefly once again.

According to the Standard Model, there are two types of fundamental particles.

o FERMIONS - Creates all types of matter.
o BOSONS - Carries all types of forces.

Let's talk about Fermions:
There are two types of Fermions:

- QUARK
- LEPTONS

FERMIONS	matter constituents spin = 1/2, 3/2, 5/2, ...				
Leptons spin = 1/2			**Quarks** spin = 1/2		
Flavor	Mass GeV/c²	Electric charge	Flavor	Approx. Mass GeV/c²	Electric charge
ν_e electron neutrino	$<1\times10^{-8}$	0	u up	0.003	2/3
e electron	0.000511	−1	d down	0.006	−1/3
ν_μ muon neutrino	<0.0002	0	c charm	1.3	2/3
μ muon	0.106	−1	s strange	0.1	−1/3
ν_τ tau neutrino	<0.02	0	t top	175	2/3
τ tau	1.7771	−1	b bottom	4.3	−1/3

QUARKS:

As you can see in the above figure, there are six types of quarks, also known as Quarks' flavor. Like other fundamental particles, they do not contain any internal structure as they are just like a tiny dot.

However, they have a special property that they have never been observed independently.

They always create a composite particle by combining with various configurations.

That composite particle is called Hadron.

The information we have about the quarks to date is because of observing the Hadron.

The most stable hadrons, as you all aware of are Proton and neutron.

LEPTONS:

Leptons are the elementary particles with ½ integer spin, which never takes part in strong interactions.

They do not contain any internal structure like quarks.

Unlike quarks, they do not need to create any composite particle for observation; they have always been observed independently.

One of the most popular Lepton, as we all aware of is Electron. There are six types of Leptons, from which three are electrically charged, and three doesn't have any charge.

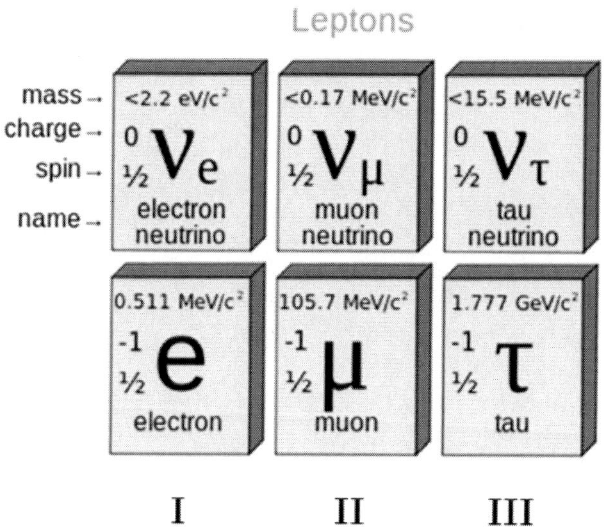

Electrically Charged Leptons are:

- o Electron
- o Muon
- o Tau

We all are aware of the Electron.

3.2 - The Electron Family

Muon and Tau are the new names for this topic. These two particles have the same charge as Electron; however, they are not called Electron because they have a bigger mass than electrons.

Let's understand it using the example below.

In the above picture, consider:

- o Cat = Electron
- o Tiger = Muon
- o Lion = Tau

Cat, tiger, and lion are the predators from the same family with different masses.

Since Muon and Tau are not stable like electrons due to their bigger mass, they immediately decay and convert into low mass leptons.

Taking Muon as an example:

Its mass is 200 times bigger than the electrons, due to which its average lifespan is only two-millionth of a second. This means that after two-millionths of a second, they get decay and convert into Electron, Neutrino, and antineutrino.

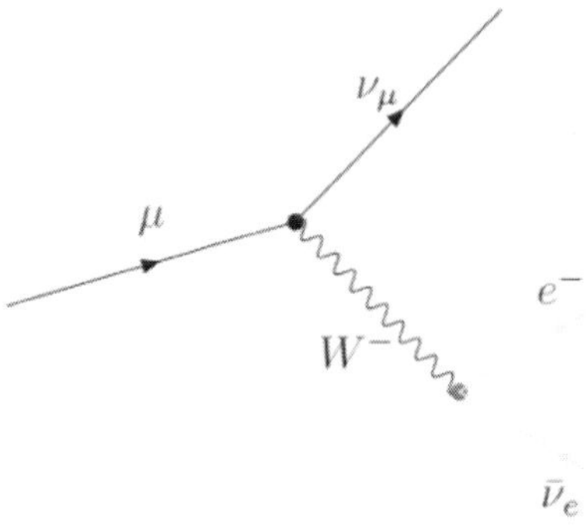

Tau is bigger in mass as compared to Muon, that is why its average lifespan is lesser. Electron, Muon, and Tau are charged particles and are easily detectable because they are easily affected by electromagnetic force.

Now, let's come to the Leptons, which doesn't contain any electrical charge. These neutral leptons are called Neutrinos.

They are not affected by any electromagnetic force and are not detectable like electrons, as they are electrically neutral. They are only affected by weak force and gravity. Since these forces are so weak, that it's difficult to detect these forces' effect on the atomic level. Therefore, it's also tough to detect the Neutrino hence named, "The Ghost Particle."

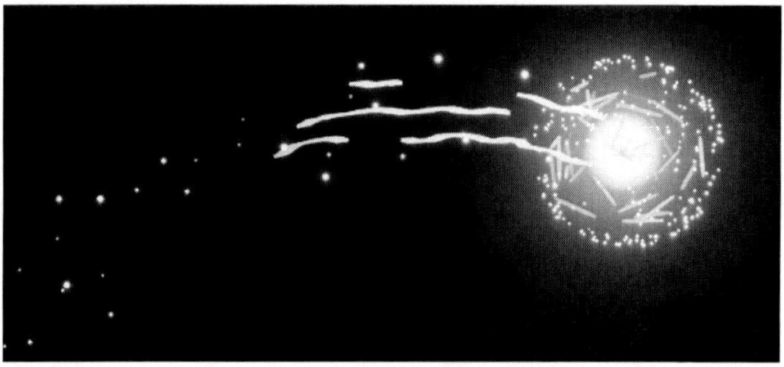

Because of less mass of Neutrino, scientists first think that they are massless like Photons. As they have very little mass, Neutrinos travels at the speed of light.

There is an interesting story about the Neutrino.

A few years back, a group of physicists researched Neutrino in OPERA PARTICLE DETECTOR (ITALY). According to their calculations, they announced that Neutrino could travel faster than the light. This news became very famous all over the world as it belies the Einstein's theory of relativity.

As the theory of relativity indicates that, an object can travel maximum at the speed of light, therefore no object can travel faster than light.

Then, five teams of researchers repeatedly did the same experiment to verify the speed of the Neutrino.

They found that Neutrino's Speed is not higher than the light, perhaps it is slightly lower. As a result of this, the physicists team of the Opera Particle Detector acknowledged their mistakes in their calculations and apologized.

Back to the topic.

There are three types of Neutrinos:

1. Electron Neutrino
2. Muon Neutrino
3. Tau Neutrino

As you can see, their names are associated with the charged Leptons.

Why so?

To understand this, we first need to understand that all the Leptons have their anti-particles as well. A mass of a lepton particle and its anti-particle is the same, but the charge is always the opposite.

For example, the anti-particle of Electron is Positron.

Its mass is equal to Electron; however, the charge is positive in comparison to Electron.

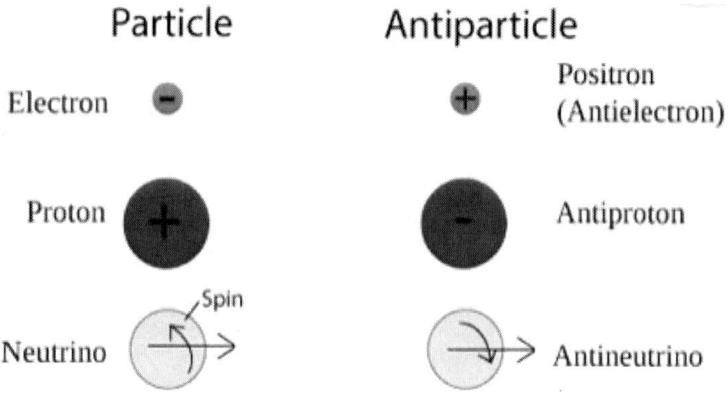

According to some physicists and theorists, Neutrino is the anti-particle of itself. This theory is correct or not; it's not yet proven.

Coming back to our question, that why are their names associated with the charged Leptons?

It is because we only get those charged particles or their anti-particles along with Neutrino during beta decay.

To understand this process, we need to understand how weak nuclear force affects the Neutrino.

An excellent example of this is when weak nuclear force interacts with W particle.

Again, you have a question that what is W particle?

For now, keep in mind that W particle contains very high mass due to which it decays immediately after its formation. (we will discuss in-depth about W PARTICLE in any other part)

Scientists combined the name of that charged particle or its anti-matter with the associated Neutrino, which they detect during the decay.

For Example, if they get Muon or Anti Muon during the decay of W particle. Then the Neutrino will be named as Muon Neutrino.

Similarly, if they get an Electron or Anti Electron (Positron) during the decay of W particle, then the Neutrino is called Electron Neutrino.

Physicists are not completely sure about the mass of these three neutrinos till now.

Oscillation is one of the processes to calculate the difference between mass of these 3 Neutrinos.

Following this process, physicists predicted a range in which their mass value should lie.

It's also challenging to calculate the exact mass of the Neutrino because rather than a single, it exists in two classifications, which cannot be combined.

For Example: if we divide Neutrino into two groups:

From the 1st group, which is according to its mass, we get.

- o Neutrino 1
- o Neutrino 2
- o Neutrino 3

From the 2nd group, which is called weak classification, we get.

- o Electron Neutrino
- o Muon Neutrino
- o Tau Neutrino

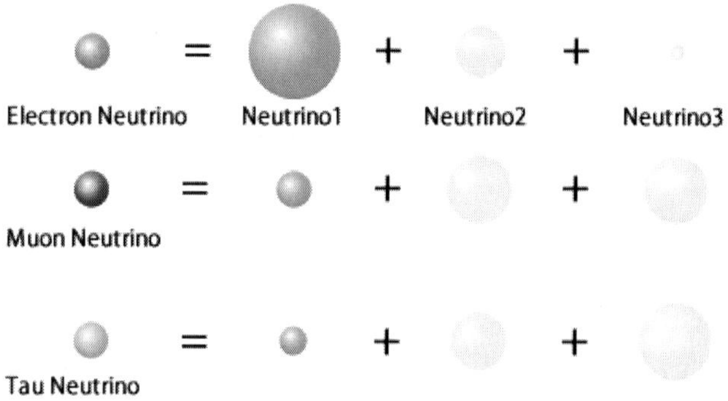

Now, when we combine the 1st and 2nd group and make the subgroups, we will get nine different combinations.

These nine combined names consist of its name and its mass.

For example:
Neutrino 1 and Electron Neutrino
Neutrino 2 and Muon Neutrino
Neutrino 3 and Tau Neutrino

These combinations makes it very easy to understand the mass of a Neutrino if the name is given and vice versa.

But due to the weirdness of quantum mechanics, we cannot calculate Neutrino by this process.

The reason is that the mass of Neutrino is not already stable like any other matter.

However, it is formed according to the superposition of all three weak neutrons.

Similarly, all weak type neutrinos are formed according to the superposition of the mass type neutrinos.

You might want to ask that what is superposition?

Superposition is the ability of a quantum system, where multiple states take place at the same time.

(We will go through "Super Position" thoroughly in the next part for Quantum Mechanics)

Due to the property of superposition, if we somehow know the Neutrino's mass, still, we can't tell with certainty about the name of the Neutrino and vice versa.

Due to the same property, a neutrino converts into another neutrino by its own.

3.3 - The Creation

Neutrinos are very difficult to detect due to its low mass being electrically neutral. They can able to pass through any object without any difficulties, whether its stainless steel.

There is a very good example here about the property of Neutrino.

Due to the nuclear reaction in the core of the sun, the production of Photons and Neutrinos takes place.

Photons interact with the sun's dense matter; that's why it takes 1000 years to get them to the sun's surface.

However, neutrinos don't interact with anything there; it comes out to the sun's surface within a few seconds.

You didn't even realize that every second, approximately 100 trillion neutrinos get across from your body.

Now the question is, what is the origin of Neutrinos?

The first discovered neutrons and recent results reveal that it came after 10^-4 seconds of the Big Bang. After a second of a Big Bang, our universe became a place where Neutrinos can travel easily.

They have always been traveling in the universe because they hardly interact with anything.

Apart from the Big Bang, they are produced in different things in different forms.

For example, they are produced by the sun; they are produced during the radioactive decay; they form by a nuclear reaction and are found in the nuclear reactor.

Similarly, they are produced when a supernova explodes; and forms when cosmic rays collide with an atom. But when they are not detectable at all, how scientists found them?

Well, there is another interesting story behind the discovery of Neutrinos.

Neutrino is experimentally detected in the year 1956, but the prediction of its existence has been already made in the year 1930.

Scientists have noticed a very uncommon thing while understanding the process of Beta Decay. If you don't know about beta decay, I would like to explain it here briefly.

BETA DECAY is a radioactive process in which a beta particle is emitted from an atom's nucleus, raising the atomic number of the atom by one if the particle is negatively charged, lowering it by one if positively charged.

Ideally, when a particle creates two new particles disintegrating itself, then the particle's energy will be the same as the combined energy of two new particles.

However, scientists have noticed that, during Beta decay, when a neutron decays and is converted into a proton inside the nucleus, only an electron is emitted.

And the weird thing was that its energy was not equal to the neutron presented inside the nucleus before the beta decay process.

The electron's energy was lower than the neutron, which means some of the Electron's energy was missing, violating the law of conservation of energy.

The scientists were unable to understand these phenomena and were also unable to find the missing energy.

There was another thing as well, which put the scientists in more doubts.

It seems like this event was also violating the law of conservation of angular momentum.

3.4 - The Anomaly

To understand this anomaly of the beta decay process, Mr. Wolfgang Pauli had first predicted the existence of Neutrino in the year 1930.

He mentioned in his theory that, during the beta decay process, Electron is not the only one that emits from the nucleus.

But there should be one more mysterious particle that existed with Electron during the emission process, which is not detectable at all.

That mysterious particle should be invisible and without any charge.

Its mass should not be bigger than the 100th part of the mass of the Proton.

This invisible particle carries that energy and angular momentum, which was missing in the beta decay process's previous calculations.

After writing this theory, Mr. Wolfgang Pauli wrote the words below:

"I have done a terrible thing; I have postulated a particle that cannot be detected."

Everyone was very convinced by the theory given by Mr. Pauli about that mysterious particle because it was completely explaining the anomaly of the beta decay process.

At first, this mysterious particle is named Neutron because there is already a particle with the same name, scientists changed it to NEUTRINO.

Neutrino has been only kept in theories for the next 26 years as it wasn't detected due to its elusive behavior.

3.5 - It Exists

In 1956, Neutrino was first detected with an experiment, and it was clear that this particle was not just a theory, and it existed.

According to the theory, beta decay should be the main source of Neutrino.

Following this, a team of American scientists successfully detected this particle in a nuclear weapons laboratory in South Carolina. They detected an Electron Neutrino with the help of two big water tanks containing liquid scintillators in it.

They detected Electron Neutrino with the help of two big water tanks containing liquid scintillators in it. This was a big invention, and Nobel Prize awarded to the whole team in the year 1995.

Until that time, everyone thinks that there is only one form of Neutrinos. However, gradually they discovered that there are three types of Neutrino.

In the 1960s, Mr. Jack Steinberger, Mr. Mel Schwartz, and Mr. Leon Lederman discovered Muon Neutrinos, and this team also got the Nobel prize in 1988.

At last, Tau Neutrino has been discovered in the year 2000. Till the time, scientists believe that there are only three types of Neutrino existing in the universe. However, no one is sure about the future. Maybe they will find something which could be very far from the expectations.

3.6 - Super-Kamiokande

A scientist in Japan has created a detector to detect many Neutrinos, which is known as Super Kamio-Kande.

It's an underground base, which is situated 3300 ft below a zinc mine. Its detector contains 50000 tons of water, which is filled in a dome-shaped tank.

There are 13000 light sensors equipped in its walls.

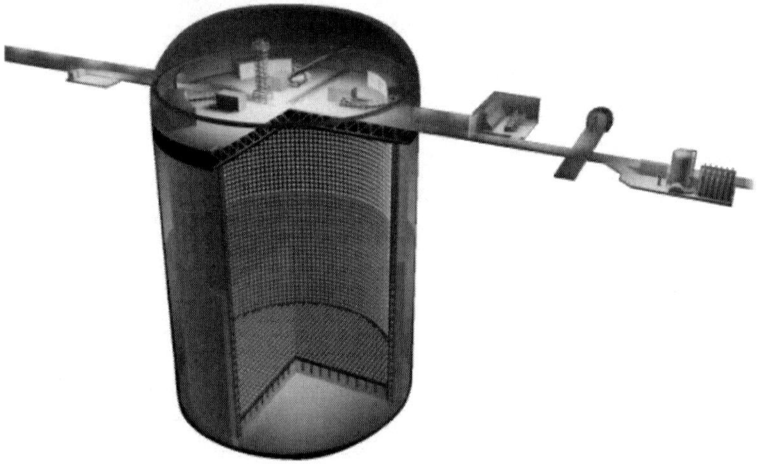

In the inner walls of the detector, approximately 12000 photomultiplier tubes of 50 CMS of diameter are fixed, which works as a light collector.

Apart from this inner detector, its outer side is also covered by a layer of water.

This outer part is also known as an outer detector.

It has light sensors that detects those charged particles which enter in a central volume.

These outer detectors also work as a shield and absorb the neutrons present outside the main water tank.

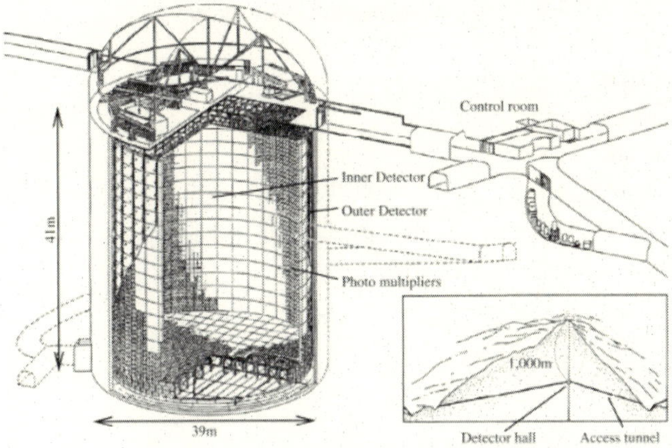

Let's try to understand how this detector works.

As it's hard to detect Neutrino directly, Super Kamio-Kande detects those by-products produced by water and Neutrino interaction.

For this, it uses a special phenomenon called CHERENKOV RADIATION.

First, let's understand what Cherenkov Radiation is.

We all know that the speed of light will remain constant in a vacuum state.

However, there are other states and mediums as well, where the speed of light is not constant, but it decreases.

Water is one of those mediums.
The speed of light decreases when it passes through the water.

If we assume that,

Speed of light in a vacuum = C

Then,

Speed of light in water = 0.75 C

So, the speed of light in water is only 75% of the light speed in a vacuum. This creates an idea that if there are energy particles presented somewhere, they can travel faster than the water's light.

Cherenkov Radiation produces a charged particle-like Electron, travels at 75% of the speed of light, in an electrically polarizable medium like water, and creates a conical pattern in its traveling direction.

When this phenomenon produces Cherenkov radiation in Super Kamio Kande, we can see a blue color light in the pure water presented in the tank, which travels through this water and collides into the detectors presented in inner walls.

These detectors contain the photomultiplier tubes, which are so sensitive that, if we put them on the moon and light a candle on earth, they are still able to capture the light of that candle.

The photomultiplier tubes capture the light produced by the Cherenkov Radiation so that scientists can study it.

It captures the amount and time of the light they received.

Later, according to the photomultiplier tubes' data, the energy and starting point of the Neutrino get reconstructed, which gives us the path from the source of their direction.

Moreover, by studying the Ring Pattern Details, we also get to know the type of Neutrino.

Whether it's an Electron, Muon, or Tau?

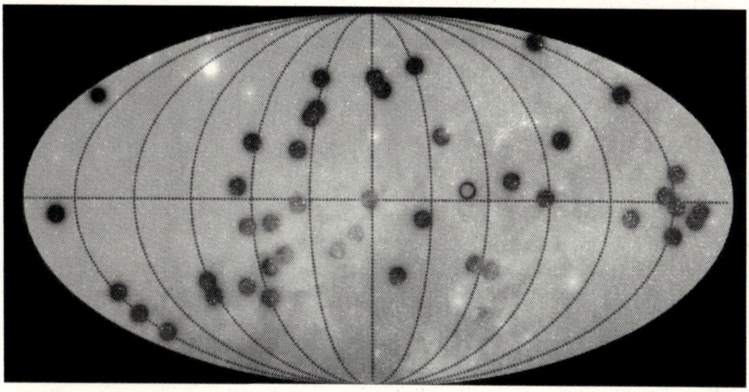

Super Kamio-Kande made a significant discovery that our sun is also producing neutrinos which are passing through all the things of the earth every day, including us.

This discovery gives excellent information about Neutrino; however, it puts us in doubt as well.

Because the value we predicted and the actual value of the neutrinos traveling to earth from the sun were different. Our detectors were detecting a smaller number of neutrinos than the predicted.

But where were the rest of the neutrinos?

This problem was named the Solar Neutrino Problem and cannot be solved for the next few years.

Finally, in 2001, scientists get to know that the missing neutrons were not actually missing.

However, they are just changing their flavor.

When we realized that neutrinos are actually capable of changing their flavor, we found the root of the problem.

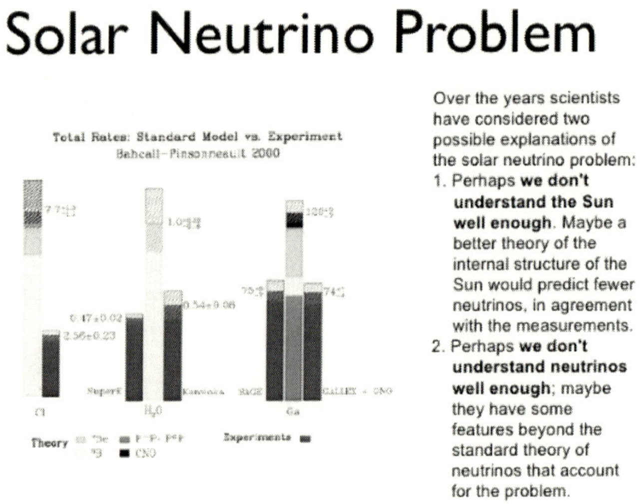

Our detectors on earth are only equipped to detect a particular kind of Neutrino, mainly Electron Neutrino.

Let's assume that 100 Neutrinos are traveling to earth from the sun.

From which 40 converted into Muons and 40 converted into Tau before reaching the earth.

In this case, we can only detect 20 Neutrinos on earth and thinks that rest 80 are missing. This changing property of Neutrino is called Neutrino Oscillation.

As we know, Neutrinos are elusive in behavior therefore they can convert into any flavor.

As neutrinos rarely interacts with any force or matter, they travel continuously since the origin of the universe.

This means that if we would be able to understand Neutrinos, we can also get to know about the universe's history, about the galaxies and other objects through which they are traveling or from where they propagate.

That is why our scientists are continuously working on detecting more and more Neutrinos and discovering new experiments to understand this particle.

As mentioned before, a big tank of water is the best tool to detect Neutrino. But we are not capable of making such a big tank.

Therefore, one Neutrino Observatory has already been is created on that part of the globe, which is totally covered with water, but in the form of ice.

Yes, we are talking about the poles of the earth.

The Neutrino Observatory at the South Pole of the earth is a proof of how eagerly scientists are working on detecting more Neutrino particles so that they can study more about this Ghost Particle.

They are also in the process of making an equipment that can detect Neutrino in space.

Let's hope that we can unlock the mystery of the Ghost Particle in the future so that we can also understand more about Dark Matter and Dark Energy.

When we realized that neutrinos are actually capable of changing their flavor, we found the root of the problem.

The Least You Need To Know

- There are six types of Leptons, from which 3 are electrical charges, and three doesn't have any charge.
- There are only three types of Neutrinos that existed in the universe.
- Muon and Tau have the same charge as electrons; however, they are not called electrons because they have a bigger mass than electrons.
- Superposition is the ability of a quantum system, where multiple states take place simultaneously.

Interesting Fact

Every atom in your body came from a star that exploded. And the atoms in your left hand probably came from a different star than your right hand. We all are star dust.

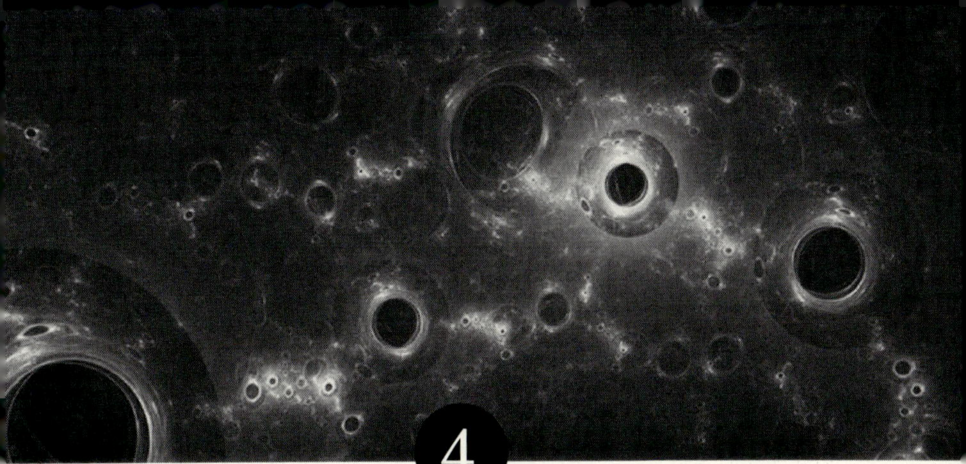

4

DARK ENERGY & DARK MATTER

To date, we were in the impression that our universe only exists in the form of molecules, atoms, galaxies, planets, trees, rocks, and fauna, which are observable.

But that is not correct.

The universe which we can observe is just a small part of the actual reality. You will be surprised to know that 96% of this universe is made up of an invisible matter.

According to the astronomers and cosmologists, our universe is 27% dark matter, 68% dark energy, and only 5% visible matter.

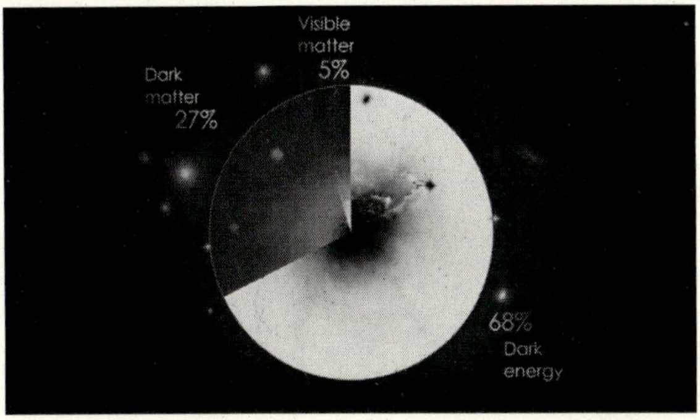

So, what is this Dark Matter and Dark Energy?

4.1 - Dark Energy Is Real

During the period of the early '90s, scientists were aware that our universe is expanding continuously.

The 2 points they were sure about the expansion of the universe are:

1. The energy density of this universe is enough that, at some point, it will stop expanding and converts into a point of singularity.
2. The energy density of this universe will be so low that it will continue expanding.

Here you might be thinking that what is Energy Density?

Energy density is the amount of energy stored in each system or region of space per unit volume.

In simple terms, the stored energy is higher if the energy density of a substance is high.

Scientists were very clear that depending upon the universe's energy density, either the universe will reach the point of singularity, or it will continue expanding.

According to the theory, it was also clear that, along with time, gravity will slow down the speed of expansion of the universe.

Scientists think this way because they believed that this universe is made up of common visible matter. The gravitational force from this matter binds the whole universe together as it is natural that the speed of expansion slows down by the attraction of gravitational force.

However, in the year 1998, the Hubble telescope clicked a picture of a supernova. Scientists discovered that the expanding speed of the universe was slower in the past than the present.

This means that the speed of expansion is increasing instead of decreasing, which was unexpected and opposite from scientists' theory.

Everyone knew that there is something very uncommon, but no one was able to understand and solve it.

To solve this weird phenomenon, they assumed energy that could be presented in the universe's empty spaces, and they named it DARK ENERGY.

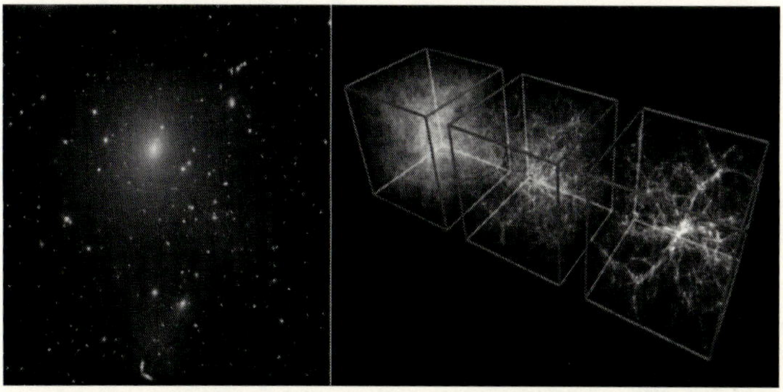

4.2 - The Emptiness Around

Dark Energy is very interesting and mysterious.

It is always presented around us and cannot be seen, calculated, or tested.

However, we can clearly see its effects.

A lot of new formations and changes are happening every second in the empty spaces of the universe.

Dark Energy is the kind of energy that can be immanent in the empty spaces of the universe because these empty spaces have a lot more energy than all other things in the universe.

There are some predictions regarding the identity of Dark Energy.

It can be one of the properties of empty spaces of the universe.

Dr. Albert Einstein once said that empty spaces are never empty; they are hiding some mind-blowing qualities.

First quality:

Empty spaces can form on their own in the universe.

And their quantity can increase along with the expansion of the universe.

Dr. Albert Einstein has mentioned about a cosmological constant in his theory of general relativity.

He also mentioned that empty spaces have their own energy, which increases along with the increasing number of empty spaces.

By this, the universe will keep expanding.

But that time, he couldn't make anyone understand the need for this cosmological constant.

Due to which everyone negated his theory.

Maybe, this Cosmological Constant is a Dark Energy.

Second quality:
These empty spaces in the universe can be sojourn and can be full of virtual particles. These particles were forming continuously by itself and were getting destroyed and invisible on its own.

The energy of these virtual particles could be Dark Energy. But here was a problem.

When a team of scientists tried to calculate the amount of energy that this energy can provide to any space, they found that their calculations were wrong. Their result is 10^120 times higher than the real Dark Energy.

The last thought is that maybe Einstein's theory of general relativity is wrong, and we need a new theory that can explain all the universe's qualities.

But what could be that new theory?

To date, Einstein's general theory of relativity gave accurate information and helped us study and understand the universe's celestial bodies.

Therefore, it is very difficult to accept and think about any kind of new theories.

So, it seems like Dark Energy remains like a mystery for now.

4.3 - Dark Matter Matters

From the last few years' research, we found that the gravity of every matter presented in the universe is not strong enough to make the galaxies and their frameworks.

If it would be strong enough, then instead of galaxies, there were stars everywhere in the universe and not their groups. This means that there is something that preserves the existence of the galaxies in the universe.

However, we cannot capture anything like this; we named it DARK MATTER.

But what is this dark matter, and what does it look like?

Unfortunately, we can't find any kind of evidence about dark matter and how it works.

To date, we only know that it existed in the universe because it acts and reacts with the gravitational force.

According to the general theory of relativity, large and heavy celestial bodies like black holes bend the space-time and light beam.

4.4 - The Cosmic Lens

Dark matter is those empty spaces in the universe that can also bend the light and prove their existence. This phenomenon is known as Gravitational Lensing.

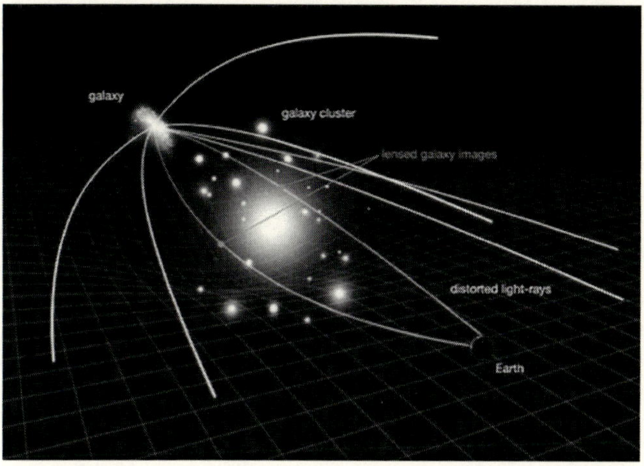

A gravitational lens can occur when a huge amount of matter, like a cluster of galaxies, creates a gravitational field that distorts and magnifies the light from distant galaxies that are behind it but in the same line of sight.

The effect is like looking through a giant magnifying glass. It allows researchers to study early galaxies' details too far away to be seen with current technology and telescopes. It is the most popular field of study in modern cosmology.

4.5 - Will We Ever Find Dark Matter?

Dark matter might be made up of strong and unusual particles, which never reacts with any kind of substance or light. We know that dark matter is not anti-matter because anti-matter reacts with similar matter and produce the gamma rays.

We are also aware that dark matter is also not made up of black holes because they exist in large numbers, and they are also not the black clouds made up of normal matter.

Then, what exactly are they made of?

Most of the scientist predicted that they might be made of Non-Baryonic Matter. All the particles like a neutron, Proton, electron, etc. are the candidates if Baryonic Matter.

According to scientists, the particle they are trying to detect is called WIMPS (weekly interacting massive particles), and its mass is 100 times bigger than the mass of Proton.

However, they don't interact with the normal particles; that is why it's very difficult to detect them.

The neutralino is also a potential candidate, which is heavier and slower than the neutron. But these particles are also not detected yet.

Some scientists believe that dark matter is a group of some invisible galaxies, and this is evident in the different laws of physics.

As mentioned earlier, we don't have enough information to understand the dark energy and dark matter completely; we still have many unanswered questions.

The theories we know about dark energy and dark matter are just the available theories.

But in the future, we might be able to dig deep into these fields.

The Least You Need To Know

- Our universe is 27% dark matter, 68% dark energy & and only 5% visible matter.
- Dark Energy is the kind of energy, which can be immanent in the empty spaces of the universe.
- Dark matter is made up of strong and unusual particle, which never reacts with any kind of substance or light.
- Dark Matter are those empty spaces in the universe which can able to bend the light and gives the proof of their existence.
- Dark matter is not anti-matter.

Interesting Fact

There is a theory which states that if ever anyone discovers exactly what the universe is for and why it is here, it will instantly disappear and be replaced by something more bizarre and inexplicable. There is another theory which stated that this has already happened.

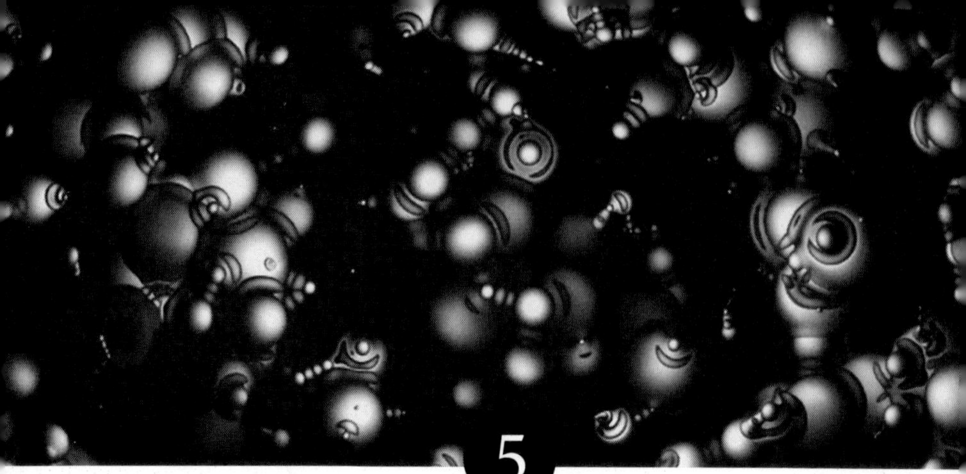

5

THE META PARTICLE

The "Tachyons" are theoretically postulated that particles which always travel faster than the light and have 'imaginary' masses.

The term comes from the Greek: ταχύ, tachy, meaning "rapid."

The complementary particle types are called luxons (which always move at the speed of light) and bradyons (which always move slower than light); both particle types exist.

Few people having some misconceptions about the Tachyon Particle like:

"These particles not only travel faster than light but also been detected already. And because these particles are detected, soon there will be possibilities of Time travel as well."

However, the reality is that Tachyon particles are hypothetical subatomic particles which only existed in theories. That means they have NEVER been detected anywhere till now. Moreover, according to scientists, there are very few chances to detect them in the future.

5.1 - The Error

Three Indian Physicists, E.C.G. Sudarshan, O.M.P. Bilaniuk, and V.K.Deshpande gave the idea of the existence of a particle that travels faster than light in the year 1962.

(1) E. C. G. Sudarshan

(2) O. M. P. Bilaniuk

(3) V. K. Deshpande

That time they named it "Meta- Particle."

After a few years, in 1967, another physicist Gerald Feinberg came up with a similar idea and released a scientific paper named "Possibility of faster than light particles."

He mentioned those hypothetical particles as "Tachyons" for the first time in his scientific papers, which means that the particles travel at high speed.

Since then, we call these particles as Tachyons.

5.2 - Law of Speed of Light

Let's try to understand the origin of the idea about Tachyons.

According to the special theory of relativity, the speed of light is the fastest one, and nothing can travel faster than the light in this universe.

BUT this is not 100% true.

Surprised?

The theory of relativity tells us that if any object's actual speed is lower than the speed of light, that object can never travel faster than light.

In other words, if the speed of any object has ever been lower than the speed of light, then it will always be lower.

If the speed of an object is equal to the speed of light, it will always be equal to it.

Similarly, if the speed of an object has ever been higher than the speed of light, it will always be higher.

If any object's speed has always been lower than the speed of light, and gradually, that object started to travel faster than light, then we say that the rule of special relativity has been violated.

But in this case, there isn't any particle that is doing the same and breaking the barriers of light speed.

Most of the particles we are aware of, travel slower than the speed of light.

That's why they can never travel equal to or faster than the speed of light.

On the other hand, massless particles like Photons, travels equal to the speed of light

That's why they have always been traveling at the same speed.

If we assume that a particle whose speed would be higher than the light speed, then it would always travel faster than the light.

We call those hypothetical particles as Tachyons.

Their existence is possible because they are not violating any rule of special relativity.

5.3 - The Way of Detection

If they exist, then what would be the main properties of Tachyons?

Let's get to know it.

Expected Properties of Tachyons:

Their rest mass, proper length, and proper lifetime will be imaginary.

According to the special theory of relativity, relativistic mass, length, and time are calculated by the three equations below.

$$m = \frac{m_o}{\sqrt{1 - \frac{v^2}{c^2}}}$$

Where
m = mass of object in motion
m_o = rest mass of object

$$l = l_o \sqrt{1 - \frac{v^2}{c^2}}$$

where

l_o = length of object in rest
l = length of object in motion

$$T = \frac{T_o}{\sqrt{1 - \frac{v^2}{c^2}}}$$

Where
T = Time in motion
T_o = Time at rest

You might be thinking that why are we concern about its relativistic mass, relativistic length, and relativistic time.

Right?

It's because an object will experience time dilation along with length contraction while traveling at high speed.

And we cannot neglect these three effects in Tachyons' case, which travels faster than the speed of light.

We take the equation of relativistic mass as an example:

$$m = \frac{m_0}{\sqrt{1 - \frac{v^2}{c^2}}}$$

m0 = Rest Mass
m = Relativistic Mass
v = Speed of the particle
c = Speed of light

Here, we are discussing Tachyons, which always travels faster than light.

That means, Speed of Particle (V) is always higher than the speed of light (C).

V>C

Hence, the value in the square root will always be negative.

We cannot calculate the proper value of a square root of a negative number. We called it an imaginary number and denoted it as it's the same as we always represent:

$$i = \sqrt{-1}$$

Relativistic mass (m) will only be real when its rest mass (m0) becomes imaginary. This means that if m0 represents an imaginary value, then all the doubts will get clear, and the equation would get solve easily.

By IMAGINARY VALUE, we meant that we could not represent it in a conventional way

OR

The mass of the Tachyons only exists when it travels faster than the speed of light.

If its speed is equal to the speed of light, due to any reason, then its mass will become zero.

Similarly, the proper length and proper lifetime of Tachyons will also be imaginary.

The speed of Tachyons increases as its energy decreases.

Usually, we see that we need to give more energy to provide more acceleration to an object.

For Example, if you want to ride a bicycle at high speed, you would have to increase the paddles' speed.

However, in the case of Tachyons, the situation would be totally opposite. It may lose energy in the form of Cherenkov Radiation. And the more they lose their energy, the more they get faster in speed.

Following this, when its energy will become zero, its speed will become infinite. Tachyon loses speed when energy is supplied to them.

Tachyons always travel faster than the light; however, if we provide it with positive energy, its speed might worsen.

They are able to reach a minimum speed equal to the speed of light. Its speed cannot go below the speed of light.

Tachyons can allow you to send messages to your past.

We would be able to send the messages through Tachyons to our past, as they can travel faster than the light.

5.4 - Possibilities of an Impossible Existence

Theoretically and by mathematical equations, there are no possible reasons to deny the existence of Tachyons. No matter how weird their properties are the mathematical equations make us believe in their existence.

However, according to most physicists, there will be no possibilities of the existence of Tachyons.

If these particles came into existence, they could violate the principle of causality and create various time travel paradoxes.

The mathematical equation of Consistent Bosonic String Theory, according to which the universe has more than 25 dimensions, also predicts Tachyons' existence.

But when we include Tachyons in this theory, it creates a fundamental instability in it.

And the solution tells us that Tachyons are always converting into the lower energy solutions after decaying.

And this cycle will always continue.

Scientists always have doubts and raise questions on Tachyons' existence because they have never been seen nor detected in any experiment.

In the year 2011, A team of scientists at CERN announced that they had observed a Neutrino which travels faster than light speed. Everyone thought that finally, they have successfully found the Tachyons.

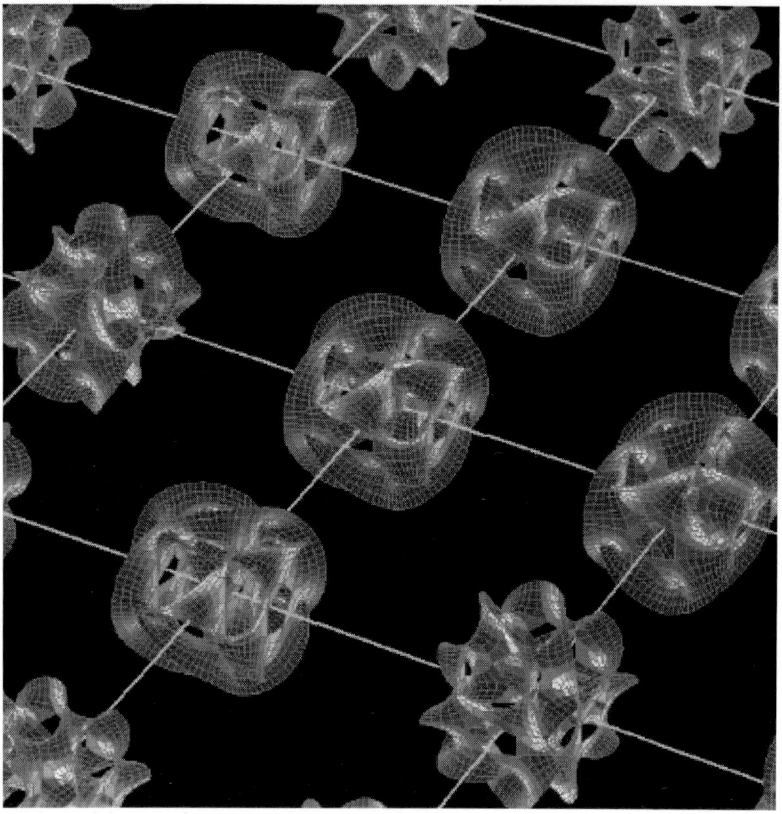

But later, they realized that it was a miscalculation by the team and what they observed were not Tachyons.

The Least You Need To Know

- Tachyons are not real, but hypothetical particles that always travel faster than the speed of light.
- They are not yet detected
- They don't violate the law of the special theory of relativity.
- Nowadays, no one mentions or discusses the existence of the meta particles.
- However, if we can successfully find them in the future, it will change the whole concept of the universe from how we know it now.
- Till then, they will remain as a hypothetical particle and remains only in theory.

Interesting Fact

Light from some stars takes so long to travel to our eyes that, when you look at the star speckled night sky, you are peering deep into the past. NASA's Hubble Telescope can look as far back as 13 billion years ago.

6

NEW LIGHT ON AN OLD NAME

Most of us know a very common PHOTON definition as "It's a packet of energy and also called QUANTA."

Suppose we ask a university student about PHOTON In that case, he will probably define the Planck's Equation of Photon Energy, which calculates the Photon energy based on the light source's wavelength.

Photon Energy : -

(Planck Equation)

$$E = hv = \frac{hc}{\lambda}$$

Where,

E = Photon Energy

h = Planck constant $= 6.6261 \times 10^{-34}$ J*S

c = Speed of light $= 3 \times 10^{8}$ m/s

λ = Photon wavelength

v = Photon frequency

But does it properly define the Photon?

NO.

While the above definition and equation might be correct in their ways; they provide very confined information to those who want to really understand Photons.

Let's try to understand from scratch what Photons are?

So, to understand Photons, we first must understand the light.

6.1 - Power of light

If you enter into a room full of light, and I ask you, ***"What's light?"*** Can you tell me what light is? No!

Imagine taking you out to a location that is dark only for a moment and return you back, then ask you the same question again. "What's light?" you'll say, "Now I know what is light because now I know what darkness is." Yet, before this comparison, you were not able to know.

In simple terms, light is the connection between us and the universe.

It is the light, due to which we can visualize and perceive everything in this universe. We can visualize the heavenly bodies, including stars, galaxies, planets, black holes & pulsars, far away from us.

Because of the light, we successfully perceived the Cosmic Microwave Background, the radiation of the afterglow of the Big Bang, or you can call it a baby universe.

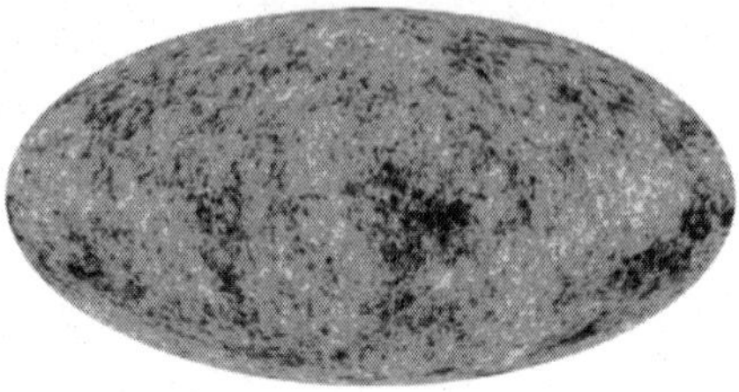

Scientists successfully found this radiation and converted it into an image.

Light also gives us information about the building blocks and elements presented in stars along with other heavenly bodies of the universe. And help us in predicting their exact position & distance from the sun.

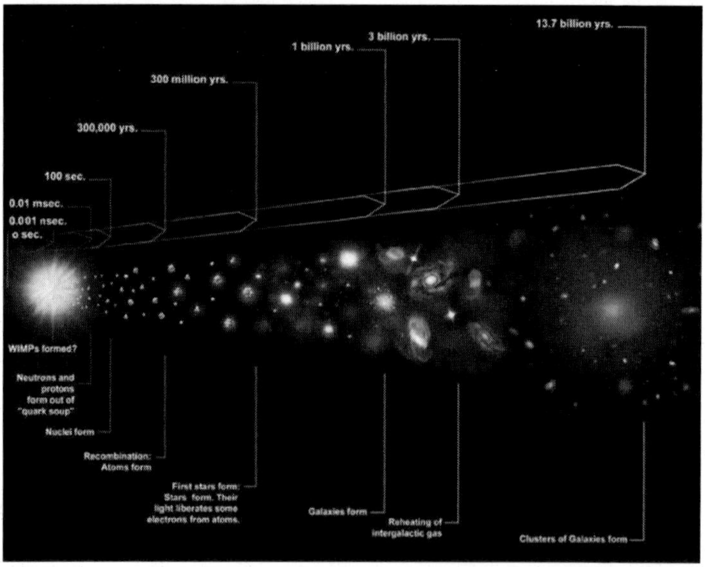

It also gives us the information about density, luminosity, and speed of celestial bodies, whether they are coming towards us or going far from us.

Light is an electromagnetic wave or radiation which travels in a vacuum from one place to another in space.

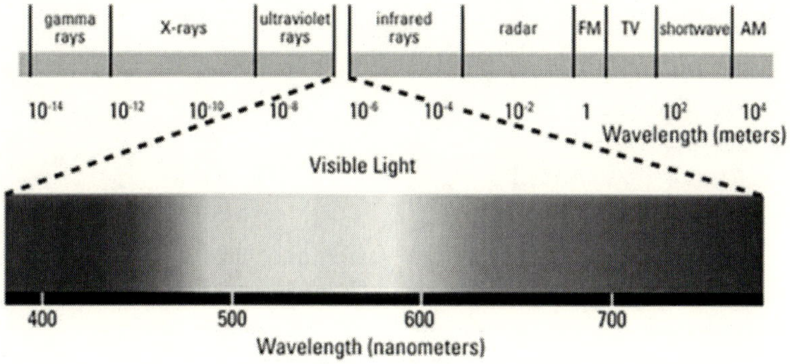

Now, how these electromagnetic waves are formed?

Let's start with an atom.

As we all know, Neutrons and Protons are presented in atom's nucleus, and Electrons revolve around its orbit. These Electrons go to an excited state from the ground state by gaining energy. Which means they jump to a higher energy level from a lower energy level.

When they move back from a higher energy level to their ground state after some time, it releases a photon.

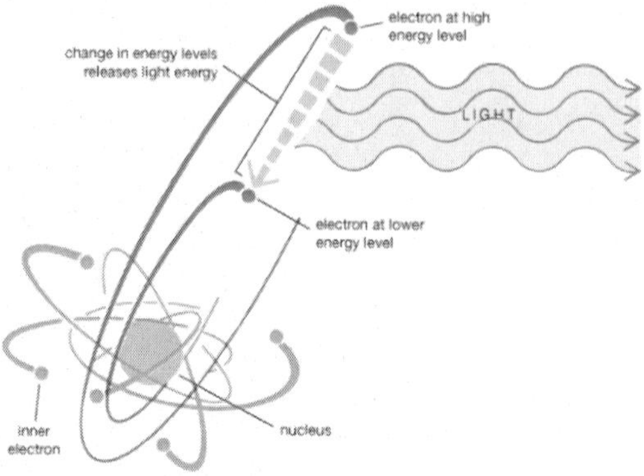

This UP & DOWN movement of Electrons give us a magnetic field.

Similarly, the movement of the magnetic field again gives the electric field.

When the electric field and magnetic field combine and oscillate at the angle of 90 degrees, that means, when they both travel in a perpendicular form, it will create an Electromagnetic Wave.

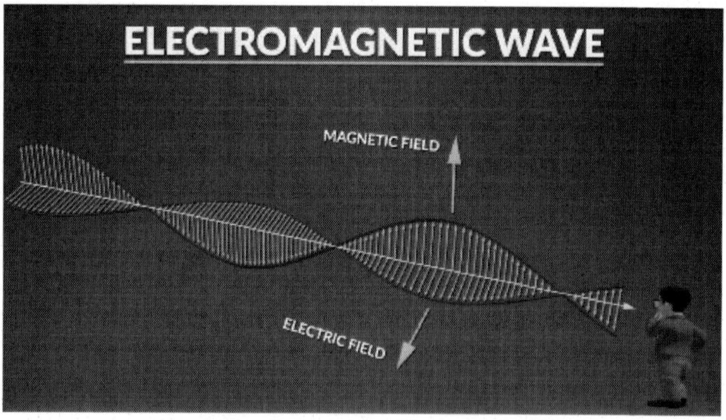

Developing a theory to explain the nature of light was a difficult task.

For a theory to be accepted in science, all the observations made on a particular phenomenon must be explained.

Light appears to have two different sets of behaviors under different circumstances. As light has a dual nature, some experiments come with result suggesting light is like a wave, while others suggested that light is a particle!

Which means, sometimes light behaves like wave-form energy, and sometimes it behaves like an extremely tiny particle. It requires the concerted effort of the best scientific minds from all over the world to develop a theory that deals with light's nature.

Let's take a look at what's meant by *"like a wave"* and *"like a particle."*

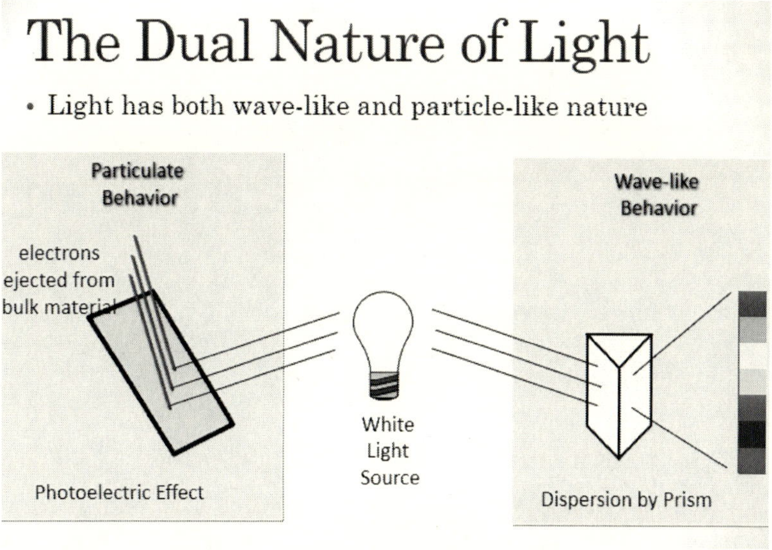

When we consider the phenomenon like interference and diffraction, we see the wave nature of light.

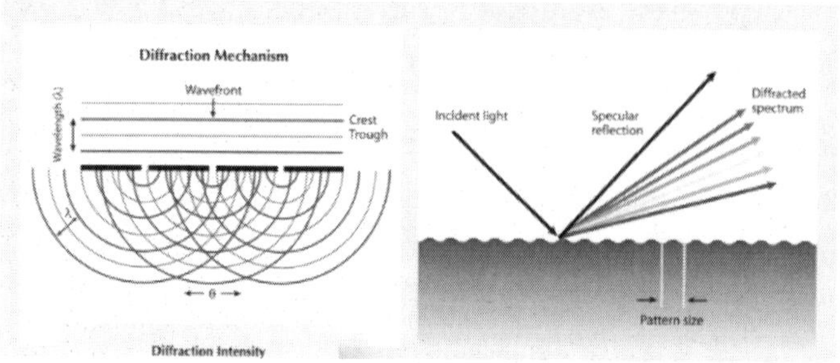

And when we see the phenomenon like photoelectric emission, we witness the particle nature of light.

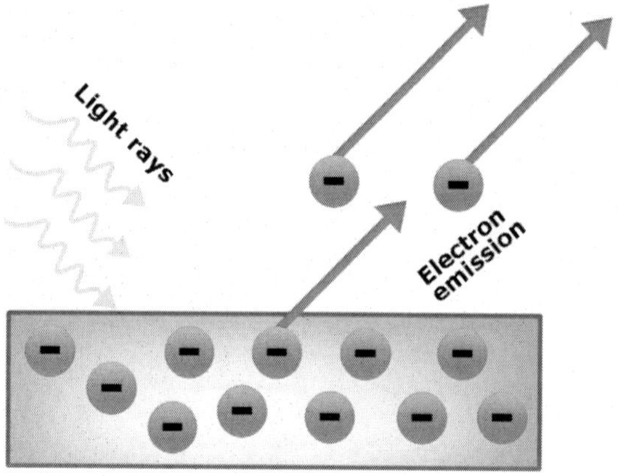

Now, we already know that light has a particle nature too.

Let's get to know this particle nature of light.

6.2 - Light as a particle

In 1901, physicist Mr. Max Planck studied the Black Body Radiation and claims that the radiation or light does not propagate in a continuous amount. Instead, it flows in the form of packets/bundles of energy. That packets of energy are called PHOTONS OR QUANTA.

A black body is an idealized physical body that absorbs all incident electromagnetic radiation, regardless of frequency or angle of incidence. A black body in thermal equilibrium (at a constant temperature) emits electromagnetic radiation called black-body radiation. This radiation has a spectrum determined by the temperature alone, not by the body's shape or composition.

According to Mr. Max Planck, every Photon carries some energy with it, which can be given by this equation.

$$E = h\nu$$

where $h = 6.6262 \times 10^{-34}$ **J•s**

v **= frequency (Hz)**

This is the only information which was given by Mr. Max Planck about Photon. He didn't provide any detailed information and described everything about Photon in this single equation.

Afterward, Mr. Albert Einstein came into the picture and gives us detailed information about the photon after studying it.

The words below are valid about the photons:

Photons are accurately and legitimately described as waves and particles at the same time. They are genuine wavicles.

You can describe Photons using different examples in different books like:

If someone throws a water balloon at you, the balloon breaks and splashes its water on you after it hits you.

OR

Suppose someone throws an egg on you. The egg gets destroyed and passes its energy to you after hitting your body.

So, there are many more examples like above to understand photon. However, every example comes up with only one outcome: photon comes in a packet of energy, passes its energy, and gets demolished.

So, let's continue with the information which Mr. Einstein gave us after studying Photons.

6.3 - Momentum of Photon

As mentioned earlier, Mr. Max Planck left Photon with just an equation. Mr. Einstein's curious brain did detailed research on it and came up with in-depth information about Photon. It heralded a new era in the world of physics.

The momentum of a photon is one of the biggest riddles that Mr. Einstein solved with his equation of relativity.

So, what is this Momentum of a photon?

We all studied the basic definition of momentum in school that momentum is equal to mass times velocity. This means momentum is directly proportional to an object's mass. Also, momentum is directly proportional to the object's velocity.

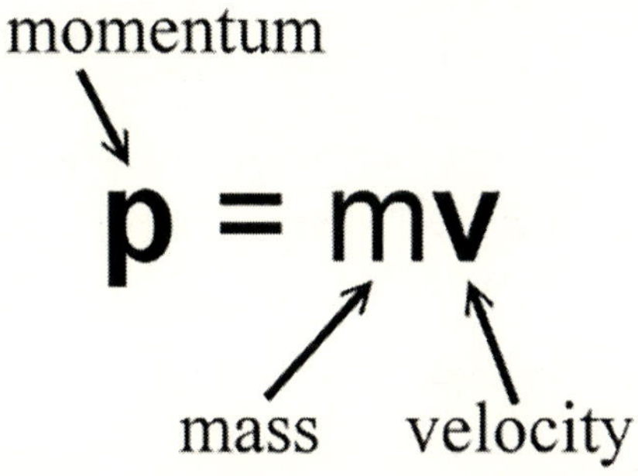

We also know that Photons are massless particles. So, following the above equation that momentum is directly proportional to the object's mass, if Photons are massless, their momentum should be zero. Right?

But this is far from the truth.

That means Photons do have momentum and contain energy too.

So, <u>the question is</u>:
How can Photons have momentum without mass?

As mentioned earlier, Photons are massless particles, but do we know that mass we are talking about here?

You might be thinking that "what kind of question is this? What did he mean by "which mass"? Mass is mass, as we all know."

But this is not the reality.

6.4 - Principle of Special Relativity

When we talk about particles like Photons that travel at very high speed, we have to apply the Principle of Special Relativity.

$$t' = \frac{t}{\sqrt{1 - \frac{v^2}{c^2}}}$$

t' Time measured from an observer outside the frame of reference.

t Time measured from an observer inside the frame of reference.

v Speed of the object.

c Speed of light

According to the Principle of Special Relativity, particles like Photons have two types of mass.

a. Rest Mass &
b. Relativistic Mass

Rest Mass: The body's mass as measured when the body is at rest relative to an observer, an inherent property of the body.

Relativistic Mass: The mass of a body in motion relative to the observer is equal to the rest mass multiplied by a greater than 1 factor, which increases as the magnitude of the velocity increases.

Generally, when we talk about the mass of any object, we always mean the rest mass of it, which always remains the same. That means the rest mass never changes.

So, when we say that "Photons are massless particles," we mean that the REST MASS of Photons is zero.

On the other hand, the RELATIVISTIC MASS of Photons is indeterminate and not mandatory to be considered because it is an attempt to resurrect the Newtonian conception of momentum as mass times velocity.

Let's say, If the velocity doubles and the momentum is MORE than doubled, then the mass must be increasing!

Right?

But it doesn't work that way.

The momentum is NOT just mass times velocity. Momentum is more fundamental than the velocity in relativistic kinematics. Velocity is an esoteric parameter and acts weird in compare to momentum.

Photons are affected by gravity, just like all other particles. "Photons fall." Their momentum is deflected by gravity. It also gets deflected when they are around any matter. They even bend space-time fabric, and all credit goes to its energy.

Let's come back to our question and try to find out its answer.

How can Photons have momentum without mass?

As per the equation of momentum (P = MV), the momentum of Photons should be zero.

Right?

But this equation of momentum is only applicable to the heavy and slow objects, not valid for particles that travel at very high speed.

$$\mathbf{p=mv}$$

This formula of momentum is only applicable to objects that are confined to only 10% of the speed of light. Hence, it will not work with a high-speed particle like Photons.

A different equation is used to calculate the momentum of objects like "Photons" in which there isn't any role of Mass. This means these kinds of objects could contain the momentum even without any mass.

Let's get to know this equation and try to understand its possibilities.

$$\mathbf{E = MC^2}$$

You all must be aware of Albert Einstein's Mass-Energy Equivalence Equation:

This equation tells us that there isn't any difference between mass and energy. This means that mass can be converted into energy, and energy can be converted into mass.

But do you know that this equation is not the complete equation, but only a special case?

The complete equation is:

$$\underset{\text{energy}}{E^2} = (\underset{\text{mass}}{m}\underset{\text{speed of light}}{c^2})^2 + (\underset{\text{momentum}}{p}\underset{\text{speed of light}}{c})^2$$

In the above equation,
E = Total energy of the particle
m = Rest Mass of the particle
c = Speed of light in vacuum
p = Momentum

This equation is derived by the relativistic definition of Energy and the Momentum of a particle and tells us that the total energy of a particle is the combination of its Mass Energy and Momentum Energy.

This also means that a particle that has Energy doesn't require Mass too in it.

Coming back to Photons, we know that its Rest Mass is zero.

What if we put this value in the above equation?

Let's have a look.

After putting m=0 in this equation, we will get:

$$E^2 = (0)c^4 + p^2c^2$$
$$E^2 = p^2c^2$$
$$E = pc$$

We can also write it as:

$$p = E/c$$

Hence, it's proved by this equation that particles like Photons don't require any Mass to have its Momentum. This is the reason that Photons has a Momentum without its Rest Mass. If this equation confused you more, and you are not convinced about the Mass and Momentum of Photons. Then let's try to understand it once again without any equation.

We already know and mentioned before that any particle requires the Mass to have a Momentum in it.

But what if it's a wave and not the particle, in that case, it didn't require the Mass.

A wave transports its Momentum in a waving motion rather than transporting any object containing Mass.

As we already know that light has a dual nature of both Particle and Wave. Here showing the wave nature, it didn't require Mass to obtain the Momentum.

6.5 - The Wave Nature

We will try to understand this Wave nature of light by a simple example:

Let's start with the Momentum:

What is Momentum?

In simple terms, momentum is the directional properties of an object in motion, which tell us how much it affects the other object if it collides or hits into it.

For Example, A car moving at very high speed accidentally collides into an another car standing at the corner of the road. In this case, we will see how much damage a moving car made to a standing car.

Talking about the Wave nature, we already know that Wave didn't require any Mass to acquire the Momentum.

For example, A huge water Wave of an ocean can damage a standing car by its Momentum without possessing any Mass.

I wanted to clarify your confusion here that water does contain Mass. However, its Waves are Massless.

I believe now you understand and get the basic concept that why Photons have the Momentum without having any Mass. If you think Photon is a particle. You are wrong. If you think Photon is a wave. You are again, wrong. The thing is that we still do not know.

The Intensity of Photons is proportional to the temperature of the body. This means your body always emits Photons at every particular temperature. Everything emits photons.

The more incredible thing is that it takes Photons millions of years to reach its surface of the Sun if it escapes from the center of the Sun. It has to pass three regions - the Gamma-ray region, the X-ray region, and the Visible ray region.

So, go outside. Look at the rays of the sunlight and get fascinated now by feeling the energy of the Photons.

The Least You Need To Know

- Photons are packet/bundle of energy.
- It can be described as a wave and a particle at the same time.
- Photons are massless particles.
- For particles like Photons which travels at very high speed, we have to apply the Principal of Special Relativity.

Interesting Fact

One year on Venus is equal to 224 Earth days. And one day on Venus is equal to 243 Earth days, which makes a day on Venus longer than a year. And to top it off, Venus is the only planet in our solar system that rotates.

7

THE ENIGMA OF ANTI-MATTER

Whenever we try to understand our universe, we always find that the whole universe comprises just two things.

First:
The objects which have mass, You, me, Earth, sun, galaxies, etc. are all Matter.

Second:
The objects which don't have any mass, but they do contain energy. We call them radiations. For example, light which travels from the sun doesn't carry any mass, but they do have energy.

Our universe is full of matter and radiation, and It wouldn't be wrong in saying that matter and radiation are the ones that define our universe. As we have proofs of it, we can see them, we can feel them, which are not just the theories. But the studies and experiments we have done so far about the universe could be wrong as well as the universe is very vast and weird than we think.

7.1 - Enter the Quantum

During the 20th Century, a great person was trying to combine the matter and radiation. And he successfully amalgamates mass and energy together.

Yes, that person was Albert Einstein. A crazy hair genius, who changed our lives by combining mass and energy by his Theory of General Relativity.

While Albert Einstein was trying to ameliorate classical physics, another community was gradually coming into the picture in that era.

On the one hand, Mr. Einstein was trying to explain the nature of the big heavenly bodies and understanding the concept of gravity.

However, on the other hand, another community was trying to explain the concepts on a quantum level. They were trying to study the behavior of atoms.

This community includes Mr. Niels Bohr and Mr. Erwin Schrodinger, explaining the subatomic particles presented in the quantized form. (see image below)

Niels Bohr and Erwin Schrödinger

Niels Bohr (left) and Erwin Schrödinger (right), along with Albert Einstein, played a role in the development of quantum mechanics, yet they were bewildered by their own theory of wave-particle duality for the electron.

Classical Physics & Quantum Mechanics both are correctly explaining their concepts.

7.2 - Un-identical Twins

In 1928, a scientist Dr. Paul Dirac came into the picture by suggesting a new theory.

According to his theory, everything presented in this universe must have its twin. He tried to explain this theory by a mathematical equation.

Let's assume that we have:

$$\mathbf{X^2 = 4}$$

After solving the above equation, we get the below two values, which both satisfies the given equation.

$$\mathbf{X = +2 \text{ or } X = -2}$$

Referring to the above example, Dr. Dirac tried to explain that everything in this universe should have its opposite.

For Example, Hydrogen must have its opposite as anti-hydrogen, our universe must have an anti-universe, and an electron must have its anti-electron.

Similarly, Matter must have its opposite called ANTI-MATTER.

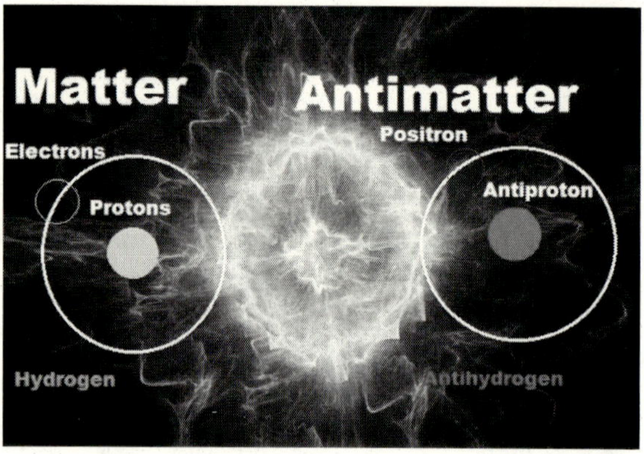

This new theory of Dr. Dirac created a fuss in physicists and scientists' community, as they were already trying to define the universe by matter and radiation only.

In 1932, Dr. Carl Anderson discovered the anti-particle of Electron and named it Positron.

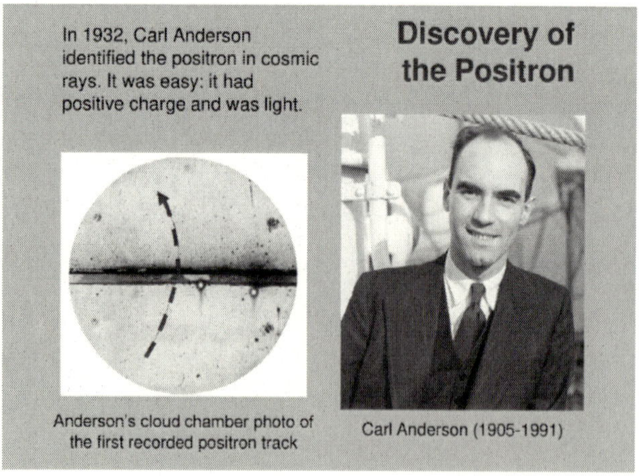

By this, it is proved that the theory of Dr. Dirac has some reasonable fidelity. This is because if positron exists, there might be other antiparticles like anti-proton, anti-atom, and anti-matter.

The word "ANTI" means the opposite. So, imagine what happens when two opposite things collide with each other?

They try to swallow each other OR in other terms, they interact with each other. What would happen when Matter and Antimatter face each other?

7.3 - The Annihilation

They would try to interact with each other and could try to fight and swallow each other. This phenomenon is known as Annihilation.

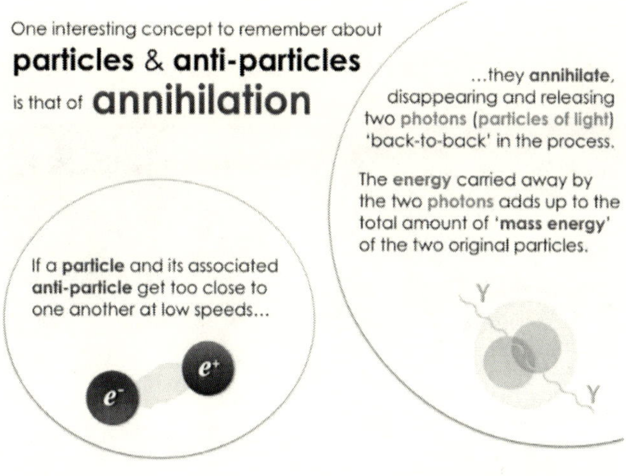

When electron and positron collide to each other and start interacting, it will release high amount of energy in the form of Gamma rays. This whole process is called Annihilation.

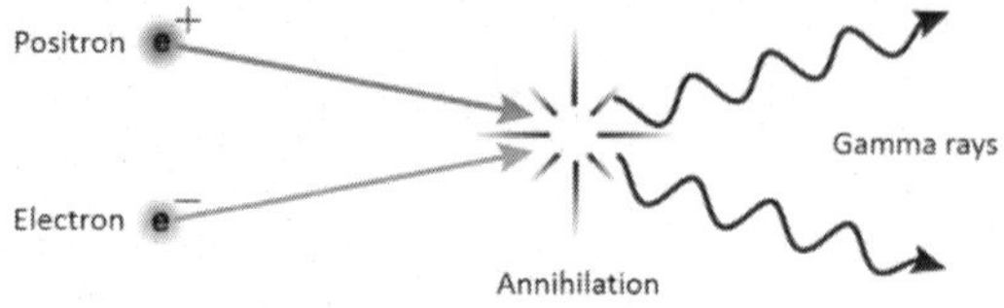

This means that Matter and Anti Matter cannot stay together, because if they will, they can emit a huge amount of energy in the form of Gamma rays due to annihilation.

Now the question is, where can we find Anti-matter?

Because if we look into the sky with our naked eyes, we can't be able to see anything.

To find the answer to this question, let's go back to that time when the universe was created.

Yes, we are talking about THE BING BANG.

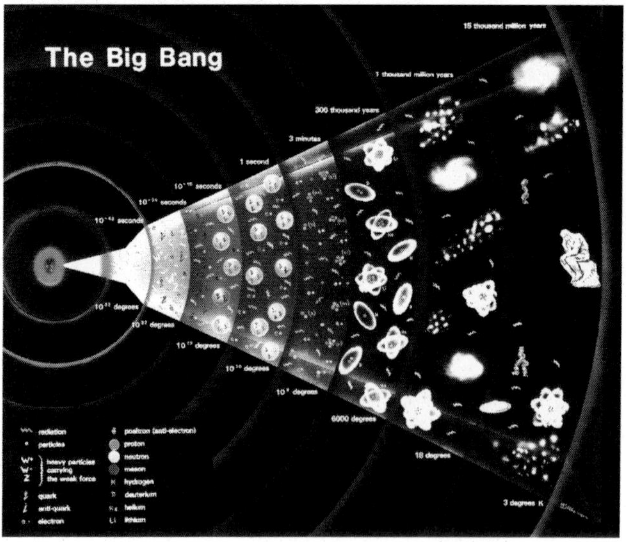

It was a huge explosion that brought the universe into existence. The tiny part of the first second after the big bang separated all the forces. It forms the fundamental particles with anti-particles like Quarks-Anti Quarks and Leptons- Anti Leptons.

When these particles and anti-particles started to form, they collided with each other to begin the annihilation process. They emitted a lot of energy in the form of gamma rays.

The same process took place with Matter and Antimatter in a very small part of a first second after Bing Bang. They annihilated to each other and released a lot of energy in the form of Gamma rays.

After this annihilation of Matter and Antimatter, some part of Matter was residuated from which our universe was created.

Now, you question why this part of Matter was residuated and did not interact with Antimatter?

Let me clear this that, after a Big Bang, the amount of Matter was more significant than the amount of its Antimatter. Because some part of Matter remains untouched, and that created our universe.

Scientists are still trying to solve this mystery and research why the amount of Matter was greater than Antimatter's amount after the Big Bang.

7.4 - Expensive Atoms of Anti-Matter

Scientists started to closely observe the universe with the help of their strong telescopes so that they can able to identify Antimatter someday. But they are still unable to find any instance related to it.

They closely monitored the Gamma radiations presented in the universe, thinking that it could be from Matter and Antimatter's annihilation. However, they cannot able to identify any big source related to Antimatter and still don't have enough information.

But, scientists are 100% sure that Antimatter exists somewhere. However, to identify it in the universe, they have to do in-depth research on it. But how would they start their research until they don't know any details on what they are doing research on?

In the end, a team of scientists decided to create Antimatter in their laboratory.

According to NASA reports, the cost of creating 1 mg of Antimatter in the laboratory is more than USD 1 Billion.

So, let's take a deep dive into the process of creating the Antimatter in the laboratory.

7.5 - The Birth

CERN is the place where scientists gave birth to Antimatter with the help of Particle Accelerator (LHC).

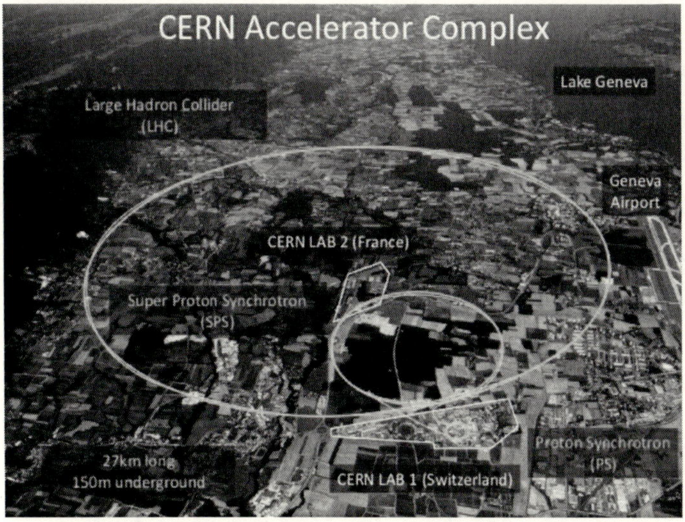

The whole process of creation is divided into two stages.

Stage One:
In the first stage, Antiproton and Positron were created

Stage Two:
In the second stage, they combine both Antiproton and Positron, which both plays opposite roles. As a result of which, we get Antimatter!

But let me elaborate this process for better understanding.

In the first place, how can Positron be obtained?

There are ***two ways of producing the Positron***

Radioactive Salt:
Positron is an anti-particle of electron and can be obtained from the β+-decay of radioactive isotopes, e.g., from 22Na (Radioactive isotope of sodium) decay reaction. In simple terms, it can be obtained by radioactive salt.

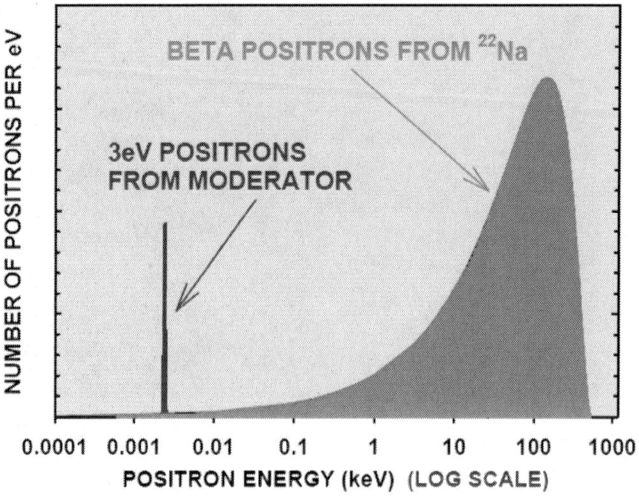

B. Pair Production:

It is a formation or materialization of two electrons, one negative and the other positive (positron), from a pulse of electromagnetic energy traveling through matter, usually in an atomic nucleus's vicinity. Pair production is a direct conversion of radiant energy to matter.

In this process, scientists make the gamma rays strike a heavy metal piece due to this a positron and an electron takes birth. From which, Positron is used to make Antimatter.

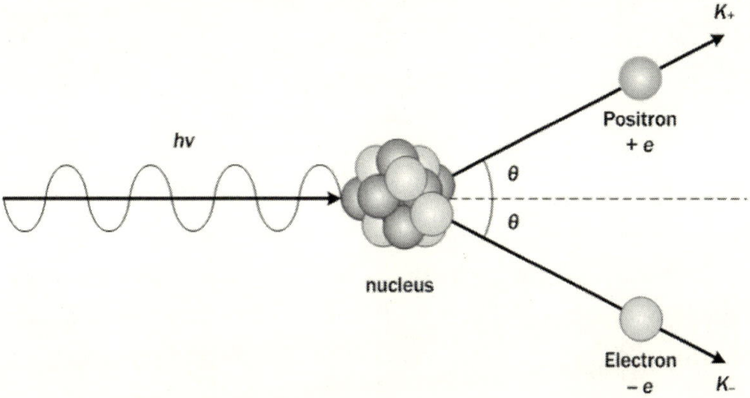

Now we have a Positron; however, we also need Antiproton to produce Antimatter.

<u>How can Antiproton be obtained?</u>

To create Antiproton, scientists use the big machine Large Hadron Collider. They use a particle accelerator to collide two protons into each other, approaching the speed of light. This collision of protons gives birth to Antiproton.

When both Antiparticles are obtained, scientists combine them and try to create an anti-atom from them. They just need to be very careful that the Antimatter they are creating should not interact with Matter. Because If it happens, then annihilation will proceed.

Now you might be thinking that there should be some process of combining these Antiparticles.

Right?

The answer is YES. There is a unique technique of combining Positron and Antiproton.

We know that Antiproton and Positron have a charge in them.

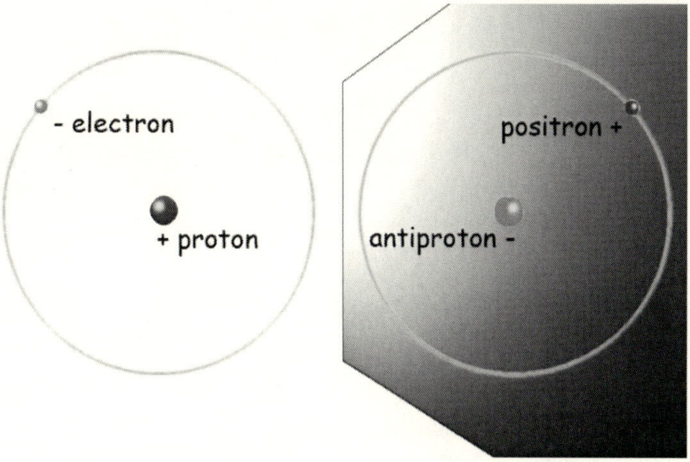

And in physics, there is a simple technique using an electromagnetic field to store the charge.

Inside a vacuum tube, scientists put Positron in an electric field and Antiproton in a magnetic field. They fluctuate the value (high and low) of the electric and magnetic field and slowly try to drag these antiparticles closer to each other. By this process, Antimatter / Antihydrogen atom comes into existence.

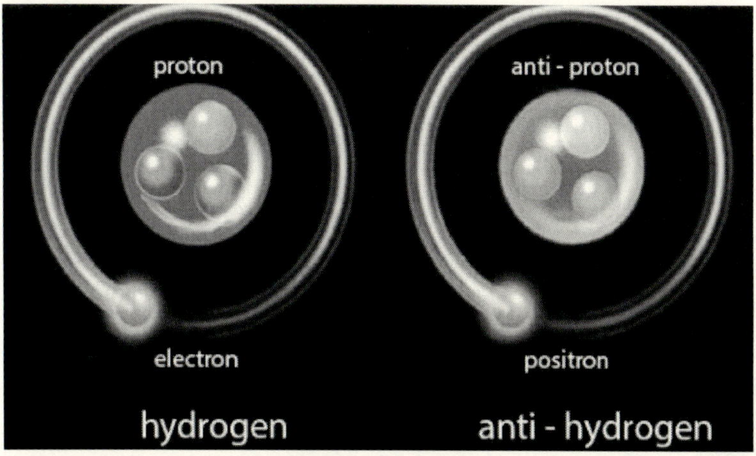

Scientists must store this Antihydrogen atom at some fixed place in the electromagnetic field. As we already know that they have to be very conscious that Antimatter should not interact with ordinary Matter. If it happens then, the annihilation process takes place and creates a high amount of energy.

Sometimes, during this process, few atoms of Antimatter start interacting with Matter. As a result, annihilation takes place and releases the energy. By this, scientists can make a confirmation that they have produced the Antimatter.

Till now, scientists at CERN have already produced approx. 300 Anti Hydrogen Atoms and stored them for approx. 16 minutes.

Why we Concretely wrote 16 minutes?

We know that 16 minutes is a short time in a normal world; however, in the Quantum World, 16 minutes are too much.

Scientists and physicists are still studying and trying to understand Antimatter by monitoring its spectrum. Antihydrogen and Hydrogen have the same spectrums, and research is still ongoing.

Antimatter could be used for producing high energy in the coming future.

If there is Antimatter's existence, then there are very high possibilities of Anti universe & Anti Earth, and they might already be created with the Antimatter.

AND, somewhere in the universe, there are even the possibilities of the existence of ANTI-YOU?

The Least You Need To Know

- Everything presented in this universe should have its opposite.

- Matter and Anti Matter cannot stay together, because if they will, the huge amount of energy releases in the form of Gamma rays due to annihilation.

- Scientists produce the Antimatter by combining Positron and Antiproton.

- 1 mg of Antimatter produced in the laboratory, costs more than $1 Billion USD.

Interesting Fact

Unlike normal matter, dark matter does not interact with the electromagnetic force. This means it does not absorb, reflect or emit light, making it extremely hard to spot. In fact, researchers have been able to infer the existence of dark matter only from the gravitational effect it seems to have in visible matter.

8

HUNT FOR GRAVITONS

In the 20th century, the two very famous theories completely changed the understanding of physics, namely:

A. Quantum Theory
B. General Theory of Relativity

8.1 - The Quantized World

Talking about Quantum Theory, it is the theoretical basis of modern physics which explains the nature and behavior of matter and energy at the atomic and subatomic levels. The nature and behavior of matter and energy at that level is sometimes referred to as quantum physics and quantum mechanics.

In simple terms, with the help of Quantum Theory, we can understand the behavior of matter.

Examples include how elementary particles can interact with each other, how they are created, and how they show their properties.

On the quantum level, the behavior of matter is always weird.

For instance, a particle can be presented in different places at the same time.

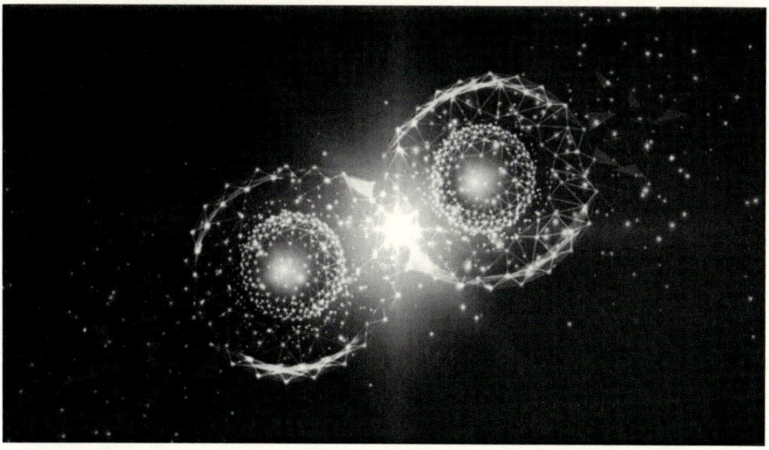

Also, a particle can be of both wave nature and particle nature at the same time.

They are also of undefined behavior, which means that we can never predict their momentum and position with 100% accuracy in a particular period.

This scenario gets a clearer picture with "The Heisenberg Uncertainty Principle," which says that we cannot measure the position (X) and the momentum (P) of a particle with absolute precision. The more accurately we know one of these values, the less accurately we know the other.

$$\Delta x \Delta p \geq \frac{\hbar}{2}$$

Mathematical representation of the Heisenberg's Uncertainty Principle

Most of the Physicists genuinely believe in this theory, as many predictions made with this theory's help are accurate.

Quantum Theory could be weird sometimes, as the quantum world just doesn't work in how the world around us works, and we don't have the concepts to deal with it. It isn't just a cool and quirky discovery, though. It's a body of knowledge that will change unexpected ways of seeing our universe and interacting with it.

Einstein was right!!!

Now, if we talk about the General Theory of Relativity, it's a theory of gravity. The basic idea is that instead of being an invisible force that attracts objects to one another, gravity is a curving or warping of space. The more massive an object, the more it warps the space around it. In simple words, it tells us what exactly is the Gravitational Force and how it works?

According to this theory, massive objects like the planet & stars warp space-time because the other objects around them start to attract towards them. This effect is called **Gravitational Force**

Most of the predictions made by this theory are resulted to be 100% accurate, which includes the very famous "Gravitational Lensing Phenomenon" and "Gravitational Waves."

Before moving forward, let me clear these two theories to you.

As explained in our previous chapters, Gravitational Lensing Phenomenon is caused by a massive body between a distant object and ourselves. It can create the appearance of two or more objects where there is only one. The light from the object gets bent around the massive body in between.

In other words, this effect can bend the path of light to such a supreme degree that background objects take on a fun-house mirror appearance, their images distorted to the point that they're almost unrecognizable.

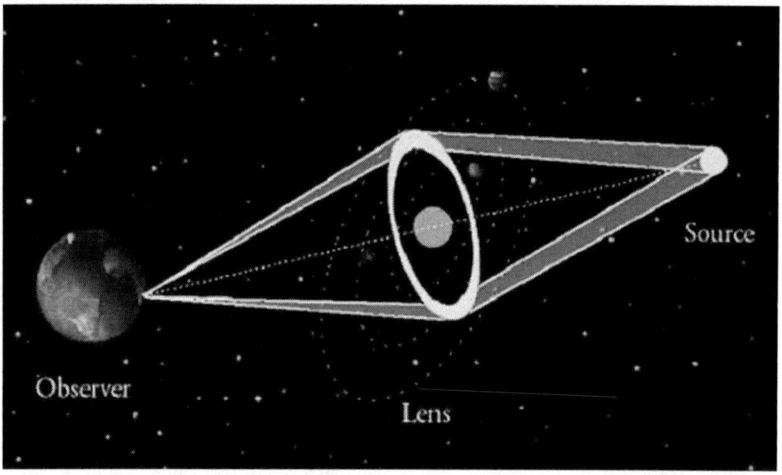

About "Gravitational Wave" it is an invisible (yet incredibly fast) ripple in space. Gravitational waves travel at the speed of light (186,000 miles per second). These waves squeeze and stretch anything in their path as they pass by. Albert Einstein predicted them in 1916, including binary star systems composed of white dwarfs, neutron stars, or black holes.

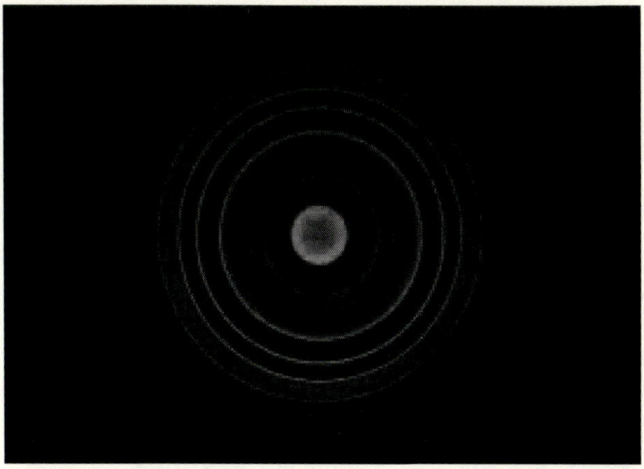

Coming back to the General Theory of Relativity, we can say that it is 100% correct in its domain.

OR

We can say that both Quantum Theory and General Theory of Relativity works accurately in their domains.

However, whenever we try to apply both principles concurrently, we always get a disaster.

For example, we know that, according to Quantum Theory, particles can be presented in many places at the same time. But, if we try to apply the same principle to our solar system, it would mean that our sun is also presented somewhere else as well at the same time.

If that happened, then all the living things of the Earth would not be able to survive.

We receive the same outcome whenever we apply the principles of General Theory of Relativity on to the Quantum World.

There are a lot of questions that are still unanswered because we have two different independent theories.

8.2 - Quantum Gravity

Quantum Gravity is an essential subject for the physicists nowadays because it seeks to describe gravity according to quantum mechanics principles. In other words, it attempts to explain how to combine the General Theory of Relativity and Quantum Mechanics.

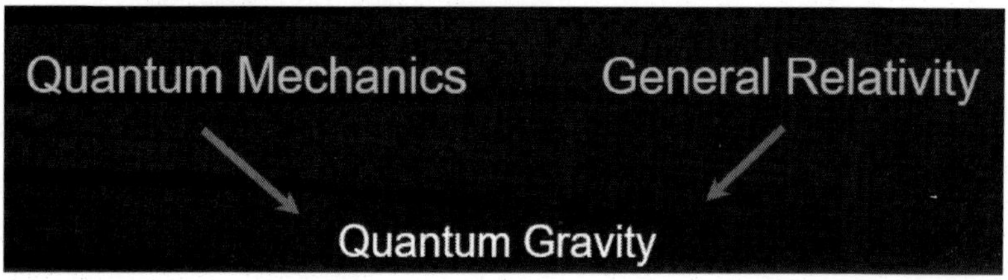

Now you might be thinking that what is this Quantum Gravity, and why do we need it?

As the name describes it, QUANTUM GRAVITY is a theory which attempts to explain the gravitational force according to the principals of Quantum Mechanics rather than classical physics. So, we can able to understand the gravitational force on a quantum level like the other three forces.

Now the next question is, why do we need to understand Quantum Gravity? There are several reasons why physicists wanted to understand Quantum Gravity.

Here we divide all the objects presented in the universe into three categories:

The first category is the objects we see and experience in our daily lives, like trees and vehicles. These are neither small like atoms nor big and heavy like planets and stars. That is why classical theory is enough to understand their behavior.

The second category is the very small objects in size like atoms, electrons, protons, quarks, etc. Quantum Theory explains their behavior in a very accurate manner. However, the only condition is that the object must be so small and light that the Gravitational Force between them can be neglected.

The third category is the objects which are very large and heavy in size, like stars, planets, black holes, etc. The General Theory of Relativity accurately explains the Gravitational Force between these large and massive objects. However, the condition is that the objects must be so large and heavy that their Quantum Properties can be neglected.

But you might want to give a thought that:

What if the object is small and heavy at the same time?

Now you must be thinking that what could be these objects, which are tiny and heavy at the same time.

The answer is: *"Singularity at The Centre of the Black Hole."*

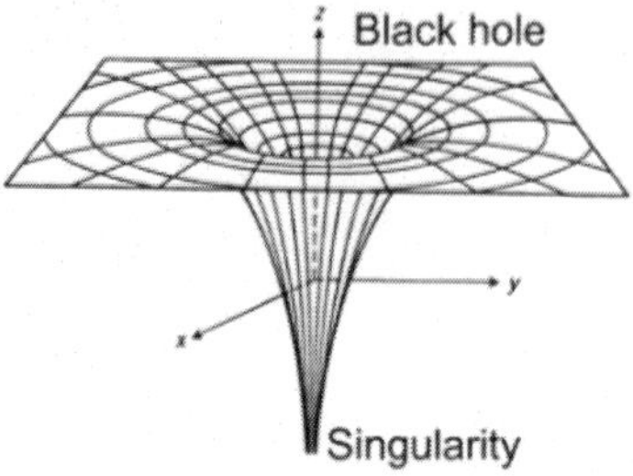

It is very small in size, however cumbersome in weight. There isn't any unified theory yet, so we are still unsure what is there at the center of a Black Hole.

Similarly, the Big Bang was started with a singularity, and we still don't have any strong theory about, what was there before the Big Bang, and why it happened?

Quantum Gravity might be an answer to all these questions!

The second reason for giving importance to Quantum Gravity is the Standard Model of Elementary Particles.

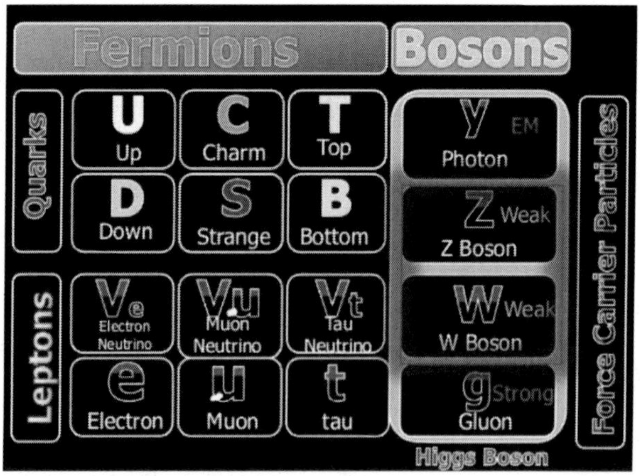

As we all know, this model has accurately explained the matter and the other three fundamental forces on Quantum Level. If this model is so accurate, then it would not be possible that only Gravitational Force could be an exception.

This clearly means that we can explain the Gravitational Force on Quantum Level!

The third reason why Physicists wanted to understand the Quantum Gravity is when physicists recreated the early universe to generate the effect of Bing Bang in Large Hadron Collider.

Then they get to know that there was a time when Electromagnetic Force and Weak Nuclear Force were not separate. Instead, they were combined as one force called ELECTROWEAK FORCE.

And before that, the Strong Nuclear Force was also not separate but combined with Electroweak Force. This means that all three forces Electromagnetic Force, Weak Nuclear Force and Strong Nuclear Force were combined together as one force in time.

If we go slightly more back in time, in that case, we will see that all four forces Electromagnetic Force, Weak nuclear Force, Strong Nuclear Force, and Gravitational Force, were combined as one Unified Force, which is later divided into two categories. Electro-Magnetic Force and Gravitational Force.

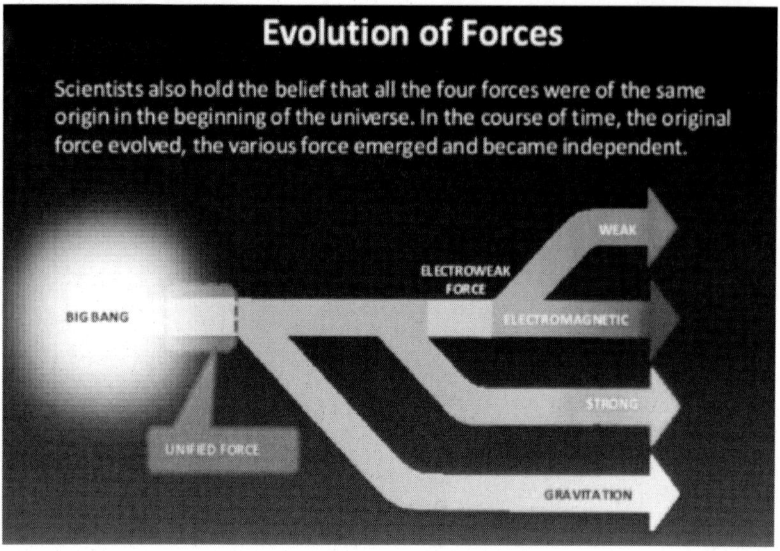

If we can find the carrier Bosons of Gravitational Force as we have for the other three forces. Then we can have one Unified Theory and, we would be able to understand and explain all kind of phenomenon and events which happens in this universe. This Unified Theory will be known as "The Theory of Everything."

If this would happen, and we successfully manage to get this theory, we can solve all the mysteries of the universe.

This is the reason why Quantum Gravity is an important subject among physicists.

8.3 - The Sound of Gravitons

There is an important thing which will be needed to complete the theory of Quantum Gravity.

Let me start with a brief explanation of all the fundamental forces presented in the Standard Model of Particle Physics.

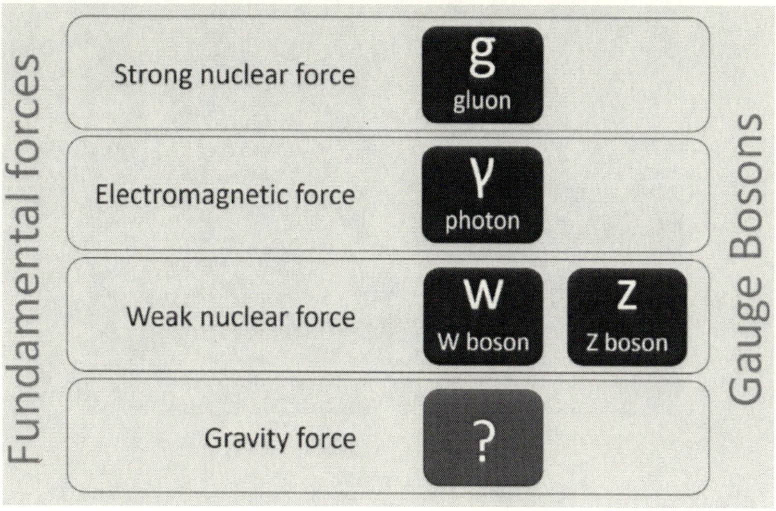

We all know that; Photons works as a carrier Boson of Electromagnetic Force. Since Photons are massless particles, their range is infinite.

Similarly, Gluon works as a carrier Boson of Strong Nuclear Force, and W/Z Bosons works as a Weak Nuclear Force carrier.

If we talk about the 4th Force, Gravity, we still did not find any carrier particles yet responsible for the Gravitational Force.

However, Physicists strongly believe that, like other forces, there should be an existence of the carrier Bosons of the Gravitational Force. They should be presented somewhere in the universe. This is the reason why the Gravitational Force works.

They named that hypothetical particle as "**GRAVITON**."

Although we have not found these kinds of particles yet, however, physicists are doing various research and experiments to find them.

In the future, if we detect some new particles, then how to know that those particles are GRAVITONs or something else?

According to the nature of the Gravitational Force, Gravitons should contain certain properties.

8.4 - The Properties of Gravitons

MASSLESS: Gravitons should be massless particles. Since the range of the Gravitational Force is infinite. This is only possible when its carrier bosons would be completely massless, like Electromagnetic Force and Photons.

SPIN VALUE = 2 Gravity is a second rank tensor field; that is why its spin value should be 2. This value is unique in comparison to all other particles. As the spin value of –

Photon = 1
Electron =1/2

CHARGE=NEUTRAL: Gravitons should be electrically neutral.

You must be thinking that if scientists already have so much information about the Gravitons and also have a Large Hadron Collider machine, then why are they unable to find it yet?

The reason is, the Gravitational Force is so weak in comparison to the other three forces that its effect is almost zero on Quantum Level.

Comparing the effect of other forces with Gravitational Force on Quantum Level

(At a size of a Nucleus)

If we assume, the strength of the Weak Nuclear Force equals one (1)

Then, the strength of the Strong Nuclear Force = 100000

AND

Also, if we assume the strength of the Gravitational Force = 1

Then, the strength of the Strong Nuclear Force is around = 100 000000000000000000000000000000000000

You can now easily estimate the strength of the Gravitational Force in comparison to other forces. As it is evident from the above example that Gravitational Force is the weakest one.

This is the reason that it is challenging to produce Gravitons in a Large Hadron Collider.

If we were not aware of the Gravitational Force already, we were only dependent on the data we received from the Large Hadron Collider. Then we would be convinced that there is no such thing as Gravitational Force exists in this universe.

The reason why Gravitational force is so weak compared to all other forces is still a mystery for the Physicists.

For now, we have only one theory that can help us find the existence of Graviton if proven correct.

This theory is known as "STRING THEORY."

I will briefly explain this theory's basic idea for now and will continue in-depth knowledge in other parts of this book.

This is a theory that looks at objects from being point-based objects to a one-dimensional string. It describes how these strings propagate through space and time and interact with each other.

We generally live in a 3-dimensional world; however, this theory tells us that there are at least ten dimensions existed in this universe. As it provides a consistent description of the universe's fundamental structure, so we can also call it **"The Theory of Everything."**

If we assume that our 3-dimensional world is a thin membrane, then the elementary particles like quarks & electrons are those open strings whose roots are always attached to this membrane. This is the reason that they can't be able to travel to other dimensions.

On the other hand, Gravitons are closed strings, so they have no roots and cannot be attached. This is the reason; they can easily travel to other dimensions.

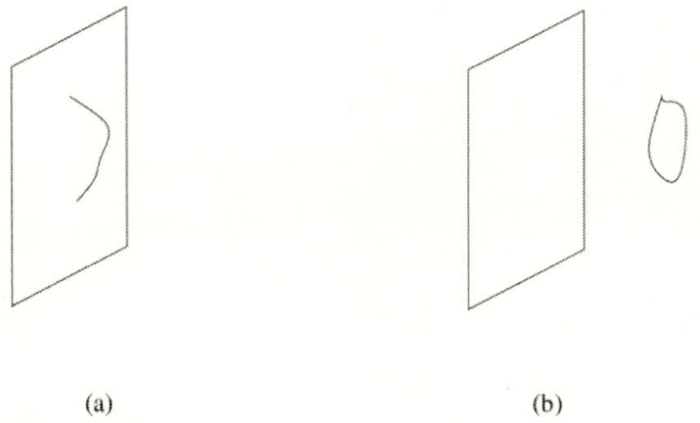

(a) open string description (b) closed string description

Gravitons are not always available in our dimension and always keep traveling to other dimensions. This is the reason that we only can experience some part of the Gravity. And this is the only reason that Gravity is so weak in comparison to other forces.

We collide the charged particles with enormous energy in Large Hadron Collider. This energy is so high that there are possibilities to produce Gravitons from it. If it happens, then we can able to find Gravitons in our 3-Dimensional world till they will not vanish into any other dimensions.

String Theory is not the only one that can explain the concept of Quantum Gravity. The major progress made in this field consists of many other theories. Below are some of the major theories which can explain Quantum Gravity:

8.5 - Major Quantum Gravity Theories

Major quantum gravity theories include -

- String Theory or Super String Theory
- Loop Quantum Gravity Theory
- Euclidean Quantum Gravity
- Super Gravity Theory
- Twister Theory
- Noncommutative Geometry
- Wheeler De Wit Equation

Super String Theory & Loop Quantum Gravity Theory are the only two of the most popular among the physicists. Loop Quantum Gravity Theory is considered as a strong competitor of the Super String Theory.

Loop Quantum Gravity

Loop Quantum Gravity has a very different methodology from String Theory to combine Quantum Theory and General Theory of Relativity. Where String Theory mentions explaining the Gravitons, Loop Quantum Gravity tries to quantize space itself — in other words; it treats space like it comes in small chunks.

Loop Quantum Gravity

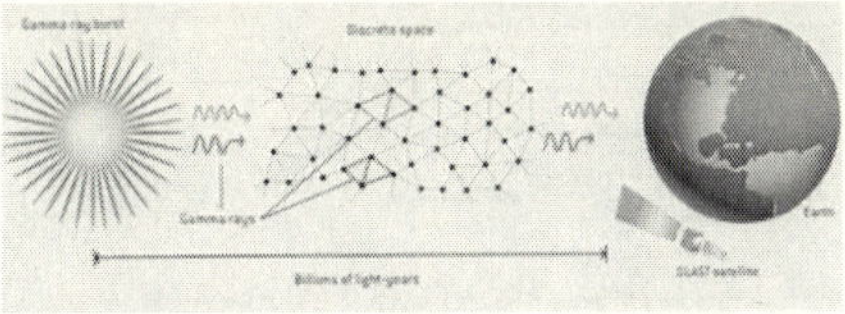

- Attempts to reconciles QM, GR
- Alternative to string theory
- Quantum space is granular, 10^{-35} meters
- Spacetime links connecting these chunks

Albert Einstein explained in his General Theory of Relativity that Gravity is not a force, but it's just a space-time property. Loop Quantum Gravity follows the same principle and says that not only matter, but space-time is also discrete, which means its also divisible.

We divide the matter into many parts and can reach to its elementary particles similarly; if we start to divide space-time many times, we can get to the point where it will not be divided anymore.

Precisely saying, like a sweatshirt is made up of a finite loop of wool. Similarly, space-time is a loop of the finite network. This loop network is known as a spin network. This network always gets evolves with time and hence known as spin form. The size of this structure is equal to the Planck length, which is approx. 10^{-35}m.

Loop Quantum Gravity Theory is still under the construction phase; however, many physicists are very convinced of its principles.

I am still trying to research more about Loop Quantum Gravity Theory and give in-depth information about it in the other part of this book.

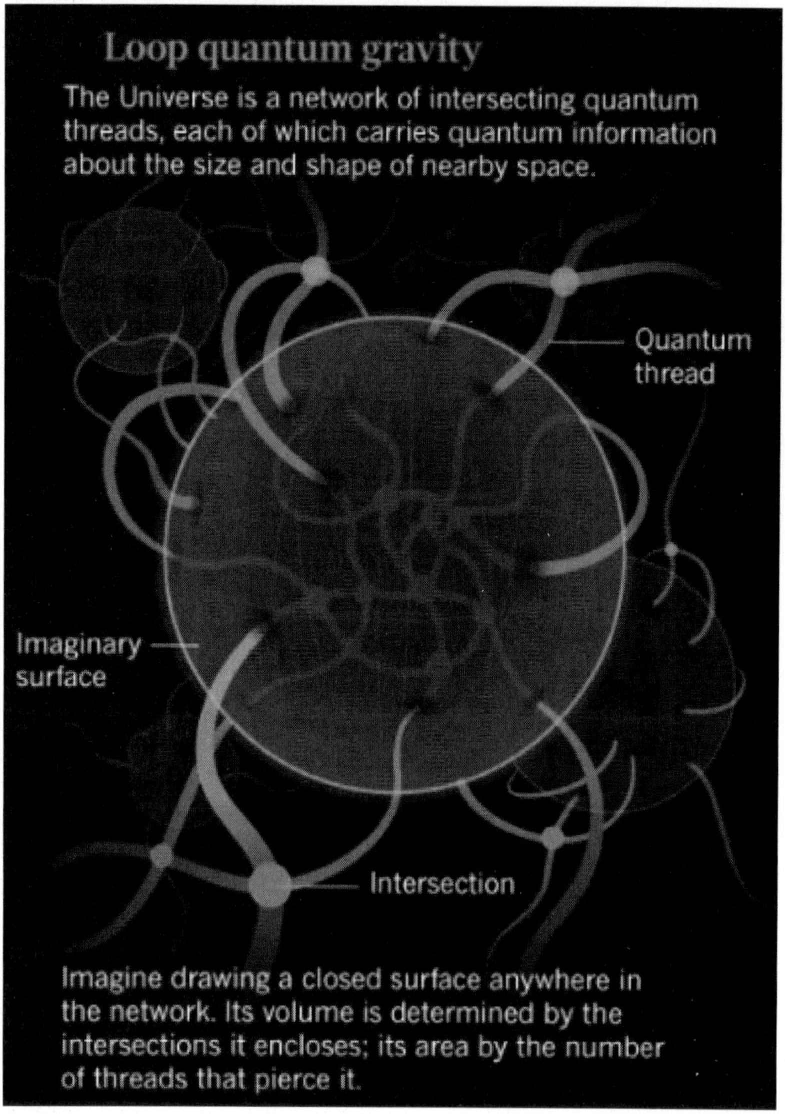

The Least You Need To Know

- Quantum Theory is the theoretical basis of modern physics that explains the nature and behavior of matter and energy on the atomic and subatomic level.
- General Theory of Relativity is the theory of gravity, which indicates that gravity is a curving or warping of space.
- Quantum Gravity is a theory which attempts to explain the gravitational force according to the principals of Quantum Mechanics.
- According to Quantum Gravity Theory, if we can be able to find Gravitons, then we would be able to understand all kind of phenomenon and events which happens in this universe.
- Loop Quantum Gravity Theory is considered as a strong competitor of the Super String Theory.

Interesting Fact

Micro Black Holes (MBH) are one of the four types of black holes, the other three being the stellar mass, the intermediate, and the supermassive black holes. Unlike the other three, MBHs are hypothetical. According to quantum mechanics, their minimum mass can be 22 micrograms. It has been said that these black holes might have played a very important role in the early universe when the energy was in the order of TeV (which is quite high).

AUTHOR' NOTE

The universe speaks to us all the time.

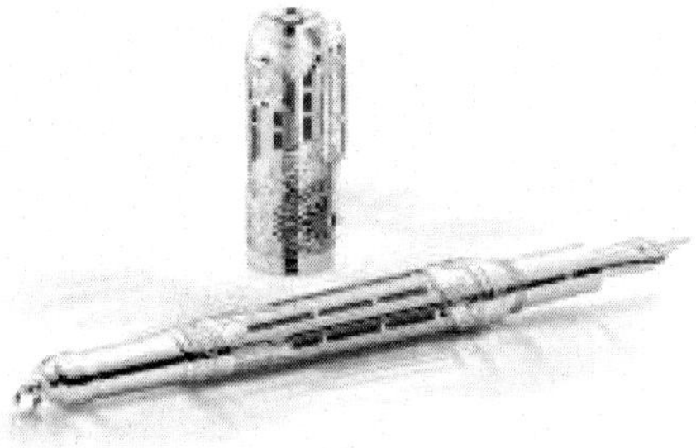

Just create a genuine desire to hear & be willing to accept the answers.

No matter how or through whom they come. It could be from a stranger, an unlikely source, or a random occurrence.

The intelligence of the universe is boundless

& the possibilities are endless.

It's always connected to you and seeking to strengthen its connection with you.

However, you must truly seek it out before it will be revealed to you.

The universe exactly knows how it must guide you.

FINAL WORD

Thank you for purchasing this book. You now have a baseline understanding of the elementary model of particles and their mechanism's fundamental concept.

If you have any direct feedback, positive and negative, or suggestions for improvement, please feel free to send me an email at – astro.physics0110@gmail.com. You can also visit my website - https://astrophysics0110.wordpress.com/

This feedback is highly valued, and I look forward to hearing from you.

Sincerely,
Deepankar Singh